SpringerBriefs in Electrical and Computer Engineering

More information about this series at http://www.springer.com/series/10059

Jinsong Han • Wei Xi • Kun Zhao • Zhiping Jiang

Device-Free Object Tracking Using Passive Tags

 Springer

Jinsong Han
Xi'an Jiaotong University
Xi'an, Shaanxi
China

Wei Xi
Department of Computer Science
and Technology
Xi'an Jiaotong University
Xi'an, Shaanxi
China

Kun Zhao
Department of Computer Science
and Technology
Xi'an Jiaotong University
Xi'an, Shaanxi
China

Zhiping Jiang
Department of Computer Science
and Technology
Xi'an Jiaotong University
Xi'an, Shaanxi
China

ISSN 2191-8112   ISSN 2191-8120 (electronic)
SpringerBriefs in Electrical and Computer Engineering
ISBN 978-3-319-12645-6   ISBN 978-3-319-12646-3 (eBook)
DOI 10.1007/978-3-319-12646-3

Library of Congress Control Number: 2014954641

Springer Cham Heidelberg New York Dordrecht London

Printed on acid-free paper

Springer is part of Springer Science+Business Media (www.springer.com)

Contents

Chapter 1
Introduction

1.1 Why Device Free Tracking?

Object tracking is one of the primary tasks of digital life and smart computing. In the past decade, device-based tracking techniques have matured enough to provide accurate location and motion information about the target in outdoor environments. Leveraging the location-aware device, such as the GPS terminal, smart phone, or RFID transponder, computer software can detect where the target is or how it is moving. With the information, people can further perceive the trace of the target, and then take specific actions.

During perception, devices embedded in the target report the location-related information, including the direct location information and indirect location information. Direct location information can show the coordinates of the target via the device, e.g. GPS readings. The latter usually contains certain physical information that is related to the actual position of the target. For example, the received radio signals strength reported by a cellphone implicitly shows the distance between it and the access point. Obviously, direct location information is more convenient for the applications. However, it requires specific location sensors, incurring a much higher cost than the indirect location information.

In outdoor environments, device-based tracking solution is widely deployed, e.g. GPS systems. The accuracy of GPS system is satisfied with the people's localization requirement. In contrast, it is still challenging to realize accurate and reliable localization techniques for indoor applications. The reasons are twofold. First, localization infrastructure is usually not available in indoor environments due the shielding effect from the build. In such a scenario, GPS signals are weak or unreachable, leading the localization serve unavailable. The cost of redeploying a new localization infrastructure, however, may be unacceptable.

Second, the majority of indoor localization solutions are also device based. Specific devices are required to be attached with or embedded in the target, which results limits or inconvenience to the user movement. In particular, the target will be uncooperative in some applications, e.g. the intruder detection. It is impossible to attach a device to the thief in advance and then detect his movements. On the

© The Author(s) 2014
J. Han et al., *Device-Free Object Tracking Using Passive Tags,* SpringerBriefs
in Electrical and Computer Engineering, DOI 10.1007/978-3-319-12646-3_1

other hand, nonintrusive sensing or detection becomes popular. With the aim of minimizing or eliminating the user intervention, the nonintrusive tracking systems are promising in everyday applications.

1.2   Why Passive RFID Tags?

Such a nonintrusive tracking system can be achieved by deploying various sensors including passive infrared (PIR) sensors, sonic sensors, and video camera sensors. Those sensors are able to provide high accuracy and sensitivity. Deploying those sensors, together with the system, incurs high cost to large scale deployments. In practice, some infrastructures, such as Wi-Fi and RFID, have been widely deployed. The wireless signals of those systems can be disturbed by the target to be tracked. Observing and analyzing the change of those signals enable low-cost and nonintrusive detection on the location and motion of the target.

In this book, we aim to design a practical device-free tracking system using RFID tags. Many RFID based motion detection or trajectory tracking approaches have been proposed in the literature. However, most existing approaches are device-based and not suitable for intrusion detection and tracking in many applications.

Our object is to reuse existing passive RFID systems, which have been deployed in modern logistic and inventory applications, for anti-intrusion. Previous works mainly use active tags to achieve this goal. However, the cost of active tags is usually hundreds times higher than that of passive tags. Although the active tag based solutions can detect the intruder with a long range while keeping him unaware, the huge deployment cost becomes a barrier of their deployments.

The solution is inspired by our observation on the interference between tags. We observed a phenomenon that when the antennas of two passive tags are approaching to each other, one tag (or both of them) becomes unreadable due to the coupling effect among passive tags. In particular, the two tags will JUST present such a phenomenon, namely critical state, in a certain distance. Keeping the distance, reader transmission power, and other factors unchanged, the one or two tags will retain the unreadable state. The critical state can be utilized to detect the movement of nearby objects. When an object moves around the two tags, termed as Twins in the following, some extra RF signals will be reflected to the tags. The one that became unreadable can harvest those signals, which will induce currents on its antenna. Because the Twins keeps a "fragile balance" between the readable and unreadable states, a small portion of energy will be sufficient to trigger the state change, i.e. from unreadable to readable state. We define this change as a state jumping. The critical state and its sensitivity to nearby movements motivate us to leverage the Twins for localization and motion detection, enabling a device-free tracking on moving objects.

1.3 What Comes Next?

Chapter 2 introduces background knowledge of RFID technology, including the brief history, terminologies, frequencies, and operation patterns. Chapter 2 also reviews the related work in the literature of localization and location based service. Chapter 3 describes the observation, operating principle, and theoretical analysis of Twins. In particular, Chapter 3 details the solution on some practical problems in system design and deployment, such as the mismatch in terms of the number of unreadable tags under critical state between the theoretical analysis and real experiments. Chapter 3 also reports the discovery of a new structure-aware coupling model between adjacent tags. Chapter 4 presents Kalman filter based and particle filter based tracking schemes for tracking the object. The two schemes are designed for dense deployments. Chapter 5 extends the Twins to sparse deployment scenarios and presents a SVD based tracking scheme. A brief Afterword rounds out the main text of the book.

Acknowledgements We would like to express our gratitude to the assistance of (in no particular order) Qian Chen, Panlong Yang, Zheng Yang, Yuan He, Longfei Shangguan, Lei Yang, Yi Guo, Han Ding, Pengfeng Zhang, Shaoping Li, Xing Wang, Wenpu Li, Wei Zhou, Yao Wei, Luyao Wang, Yingying Gou, Ge Wang, and many others.

We greatly appreciate the support, guidance and encouragement given by Springer's team, including Jennifer Malat and Susan Lagerstrom-Fife.

Chapter 2
Background

The establishment of RFID based localization requires a deep understanding on the component, operation principle, and propagation of RF signals. It is necessary to introduce the background information about RFID systems before stepping towards the RFID based localization. In addition, we will introduce the related localization approaches in the literature in this chapter.

2.1 Radio Frequency Identification

2.1.1 RFID History and Practice

The RFID technology was derived by the usage of Radar for object detection. With the increasing demand of detecting airplanes beyond visual range, radar techniques were rapidly developed in 1930s. With the backscattered microwave, the radar operator was able to alert when the aircraft moved at hundreds of kilometers per hour. However, one of critical tasks of radar based detection was not well solved at that time, distinguishing the side to which the aircraft was belonging.

The first solution was found by the German air force in a simple way. Some fighters from the German air force suddenly performed some maneuvers, for example a roll, before the air combat. Sometimes, a squadron of fighters did this behavior without any reasons. Later, this behavior was intercepted as a very naïve but effective Identification of Friend or Foe (IFF) solution. By rolling together, the Luftwaffe pilots could change the backscattered signals to the radar, such that the "modulated" signals shown on the reader screen appeared a specific pattern. In this way, the German radar operators could identify them as the friendly targets. This story is a typical case to show how passive RFID systems work via backscattered radio waves.

Considering the basic goal, the information modulated into the maneuver based behavior is too limited for identification. First, the maneuver can be performed either by friendly or foe targets, lacking of effective protection on the modulated

© The Author(s) 2014
J. Han et al., *Device-Free Object Tracking Using Passive Tags,* SpringerBriefs in Electrical and Computer Engineering, DOI 10.1007/978-3-319-12646-3_2

information. Second, the contained information by this behavior is also limited, for example, only 1 bit in a roll. But such an idea inspired the development of passive backscattering communication, which is the basis of RFID technology. The most distinct feature of passive backscattering is that the object is not with any transmitter. The object scatters back the radio signal transmitted from the radar station for identification.

After World War II, the RFID technique was not developed rapidly, due the highly expensive and large sized transponder. Later, with the development of very large scale integrated circuit (VLIC), people were able to produce extremely cheap and small chips as well as circuit components, resulting a rapid progress in the RFID manufacture. Using RFID tags to achieve more automatic and intelligent identification finds an increasingly requirements from many applications, such as the asset management, logistics, supply chain, access control, etc.

An RFID system comprises of three parts, the reader, tag, and backend server. The reader is used for read/write tags, determining the operating frequency and range. For most RFID systems, the tag is a media that stores certain information about the user or the object it attaches. The identity is the most commonly used information stored in the tag. Therefore, the identification, in which the reader retrieves the ID from a tag, is the most essential functionality of RFID systems.

Generally speaking, the major functionalities of the reader include:

• The reader communicates with the backend application via some standard interfaces. The backend application actively operates the reader for communicating with the tag, and exchanges with the reader with three major types of information: the information of tags (e.g., ID), identifier of the reader, and necessary information about the operation.
• The reader tackles the collision among tags, and identify each tag within its interrogation range. To this end, people develops the anti-collision algorithms for RFID systems.
• The reader can detect the error in the reading/writing tag procedures. Due to the ambient interference to the wireless communication between the reader and tag, the error detection and correction are very important for RFID systems to achieve reliable identification.

Normally, the backend server and the reader are jointly termed as "*reader*". Without loss of generality, we also term the combination of reader and backend applications as reader in the following. We show an example of RFID systems in Fig. 2.1.

2.1.2 Active and Passive RFID Tags

Three major types of RFID tags are currently available in the market. There are active, semi active (or semi passive), passive tags.

Active tags generate the high-frequency RF signals using the power from their own batteries. The battery also provide power to tags for their internal operations,

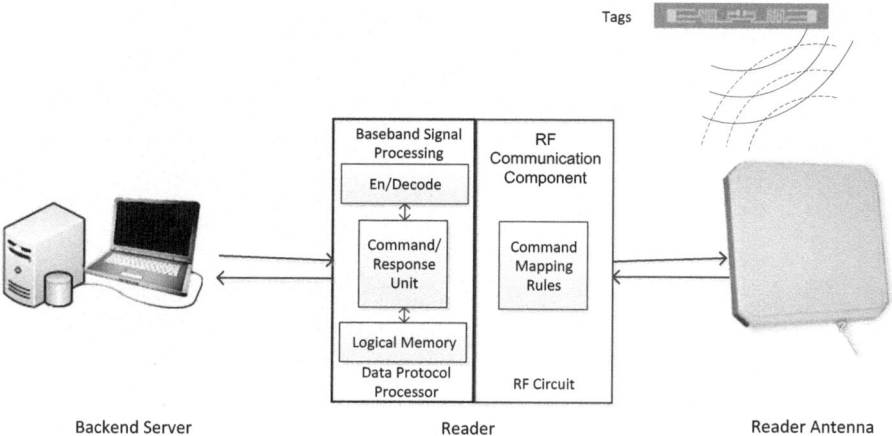

Fig. 2.1 An example of RFID Infrastructure

such as the modulation and computation. The normal operating range of active tags is often tens of meters, sometimes above 100 m. Due to limited capacity of battery, the lifetime of active tags is not long. It is non-trivial to replace the battery after deploying a active RFID system. The large size and high manufacture cost are other two issues for active RFID systems.

Passive tags are not with battery on board. They can only backscatter the RF signals sent from readers. Meanwhile, the passive tag harvests the energy from those signals for modulating its data to the backscattered signals. The interrogation range of passive tags is generally much shorter than that of active tags, e.g. ranging from only several centimeters to a few meters. On the other hand, passive tags are generally cheaper and smaller than their active counterparts, which is its major advantage for large-scale implementation.

Semi active (or semi passive) tags are a combination of active and passive RFID transponders. They use their own power for the modulation while adopting the backscattering data transmission pattern similar to passive tags.

The tag has two functional components, the antenna and chip. Figure. 2.1 shows a passive tag. The tiny rectangle in the center is its chip, while the other metal parts are its antenna. The chip is in charge of decoding/demodulating the command from the RF signal sent by the reader and encoding/modulating the response or data to the returned RF signal. As aforementioned, the active tag will actively generate the RF signal, while the passive tag will backscatter the RF signal sent from the reader. There are also two major types of antennas used by current RFID systems, coil and dipole antenna, for the inductive coupling and electromagnetic radiation systems, respectively. When the antenna is connected with the chip, it would introduce high capacitance and impedance. For achieving a good power matching, the total resistance should be equal to the maximum load resistance of antenna. On the other hand, the antenna is usually designed as a twisted shape to minimize its size Fig. 2.2.

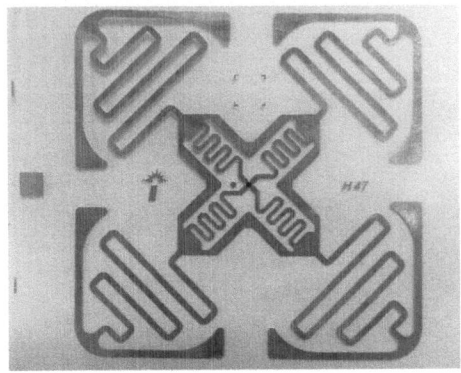

2.1.3 Frequency and Operating Range

The most important factor in passive RFID systems is the operating frequency of the reader. Although the frequency used by existing RFID systems is widely distributed, say from 135 kHz to 5.8 GHz, the available frequency spectrum for a certain type of RFID devices is limited, e.g. 860~960 MHz for UHF passive tags. There are four typical frequency ranges used in practice: 135 KHz, 13.56 MHz, UHF (433 MHz, for active tags, 860–960 MHz for passive tags), 2.45 GHz, and 5.8 GHz.

The operating patterns between the reader and tags are separated into two groups. The RFID systems that use the frequency below 30 MHz are belonging to the inductive coupling, while those that use the frequency above 30 MHz are using electromagnetic radiation. The frequency selection is generally determined by consideration of RF wave's penetration and absorption. For example, the low-frequency RFID systems are usually used for the better penetration capability. For the higher-frequency RFID systems, the operating range is larger and the sensitivity is better than the lower-frequency one. In this paper, we focus on the passive UHF RFID systems working on the 860–960 MHz spectrum, where the bottleneck of operating range is the transmission power of the reader.

2.1.4 Inductive and Radiative Coupling

Generally, the RFID systems mainly use two ways for communication, inductive computing and radiative coupling.

The inductive computing is usually adopted by the lower frequency RFID systems, such as the LF and HF RFID tags. The inductively coupled transponder uses a large-area coil or conductor loop as its antenna. The reader also has its coil for the communication. The reader actively generates an electromagnetic field using its coil. If the wavelength of the operating frequency of the reader is much larger than the distance between the transponder and reader, the electromagnetic field can be treated as a magnetic alternating field, which penetrates the area around the coil. For

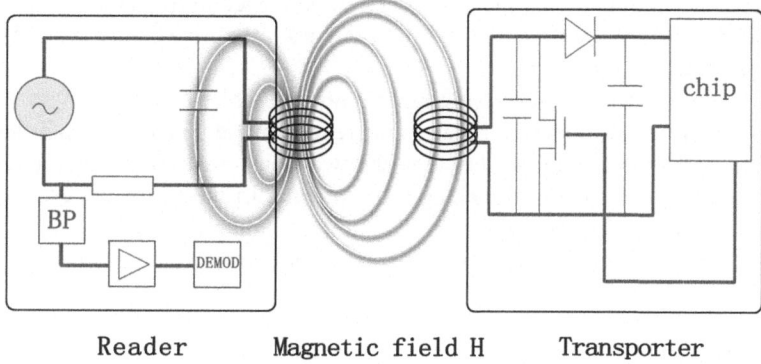

Reader   Magnetic field H   Transporter

Fig. 2.3 An example of inductive coupling systems

the part that penetrates the coil of the transponder, it induces a voltage on the transponder's antenna. In addition, the inductive coupling is actually a transformer-type coupling, which the induced voltage is rectified and used as a power supply. Since the RF signal sent from the reader is not a stable and continuous power supply, a capacitor serves as an energy bank in the transponder. The transponder then requires certain time periods to charge the capacitor for accumulating sufficient energy for operation. In practice, the tag usually switches between two states, standby and power-saving modes in the charging procedure for achieving an efficient charging (Fig. 2.3).

Besides the charging cycle, the inductive coupling is also used for data transmission. The transponder activates an on-chip oscillator, resulting in a weak magnetic alternating field. Similarly, the generated magnetic alternating field also penetrates the reader coil, inducing a weak voltage on the reader coil. The modulation can be achieved by changing the on/off state of the impendence connected to the coil such that the induced voltage shows corresponding high/low state. In this way, the data can be transmitted between the transponder and reader.

Radiative coupling, also known as the electromagnetic backscatter coupling, support a long-range communication. The antenna of radiative coupling transponders is comparable in size to the wavelength of their operating frequencies, including the UHF frequencies of 868 MHz (Europe) and 915 MHz (USA), or the microwave frequencies 2.5 and 5.8 GHz. In the radiative coupling, the reader antenna radiates continuous electromagnetic waves (CW in short). According to the free space path loss principle, the CW power fades in a relationship to the square of the distance traveled. A portion of the wave interacting the transponder antenna induced a coupled current, which can be used as the power supply. Another portion of the original CW is scattered to different directions, in which a small portion of CW will be backscattered to the reader. In this procedure, the reader can modulate the data it transfers via amplitude shift keying(ASK), frequency shift keying(FSK), or phase shift keying(PSK). For the cost concern and simplicity of demodulation, the majority of passive RFID systems use ASK modulation. On the other hand, the IC circuit of transponder can change its impendence in time with the data to be transmitted, and hence vary the relationship between the impendence of transponder Z_T and the impendence of transponder antenna Z_A. In this way, the transponder modulates its data to the backscattered CW.

Corresponding to the inductive coupling and radiative coupling, there are two types of communication ranges between the reader and tag, near-field and far-field. According to [1], the boundary between the two fields is $R=2D^2/\lambda$, where D is the size of antenna and λ is the wavelength of antenna. In the near-field communication, the interaction between the reader's and tags' antennas is based on inductive coupling [2]. Far-field communication operates based on radiative coupling.

2.1.5   Dipoles Antenna and T-Match Structure

Most passive tags use a half-wave dipole antenna. In our system, Twins uses commercial passive tags modeled as Impinj E-41b in Fig. 2.4 The length of antenna is usually half of the wavelength $\lambda/2$, i.e. 16 cm (with the operation frequency 915 MHz). The antenna is bent to form a *meandered dipole* for further reducing its size. However, meandered dipole introduces mismatching impendence between the antenna and the IC circuit in the tag, which might result in a small power transfer coefficient and inefficient energy absorption. Manufactures then adapt a short antenna to connect the capacitive IC load, forming a T-match structure, as shown in Fig. 2.5. In this structure, the impedance is balanced between the longer meandered dipole (with the length of L) and shorter dipole (with the length of a). The IC of the tag connects to the meandered diploe via two wings of the short dipole Fig. 2.6.

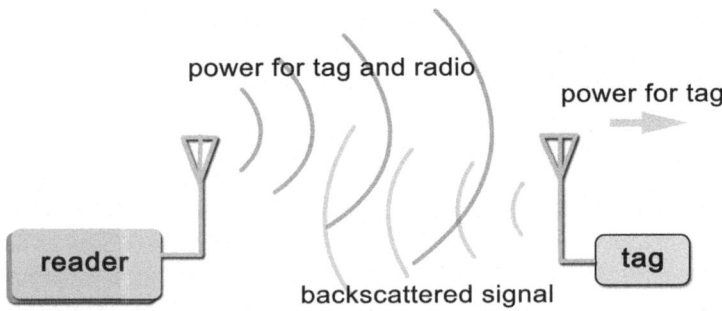

Fig. 2.4 An example of radiative/backscatter coupling systems

Fig. 2.5 Half-wave dipole antenna of passive tags

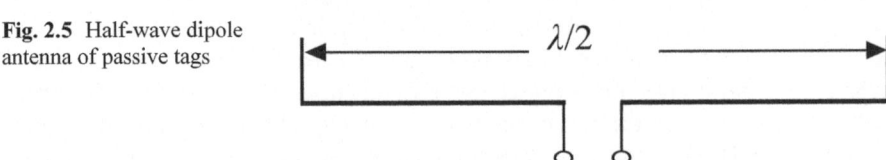

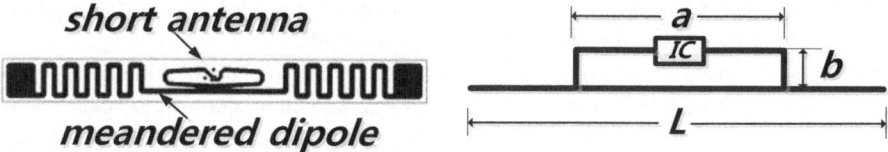

Fig. 2.6 Real tag model and T-match structure

2.2 Localization and Location Based Service

Location based services (LBSs) have been a part of our life. In the literature of indoor localization, many techniques have been proposed in the recent decades. Generally, they fall into two categories: model-based and fingerprinting-based.

2.2.1 Model-Based Indoor Localization

Model-based indoor localization approaches use geometrical models to figure out locations. In those methods, locations are calculated instead of retrieving from known reference data.

RSSI-based ranging algorithms calculate distances among nodes based on the Log-Distance Path Loss (LDPL) model according to the measured RSS values. These approaches decrease the localization accuracy due to the irregular signal propagation, especially in indoor environment. Lim et al. [3] use WiFi sniffers deployed at a known locations to measure the RSS from APs, and then use the LDPL model to build RSS map. Madigan et al. [4] use a Bayesian hierarchical model without the need of locations of the training points. However, they are still required to have prior knowledge about the AP locations. To avoid the use of AP localizations, EZ [5] builds the physical constraints of wireless signal propagation based on the LDPL model and uses the genetic algorithms to solve the positioning problems. RSSI is susceptible to multipath propagation, which results in large errors inside. Wu et al. [6] proposes to extract the dominant cluster of paths from CIR to reduce the ranging error from RSSI. They implemented a prototype, namely FILA, on OFDM based WiFi with off-the-shelf NICs, and leveraged the channel state information (CSI) to alleviate multipath effect at the receiver. Centaur [7] combines RSSI and acoustic ranging.

Other than the RSS related model, the geometric models are also exploited to characterize the relationship of signal between transmitters and receivers. PinPoint [8]and the work proposed by Werb et al. [9]use the time delay in signal propagation to estimate the distance between wireless nodes. GST [10] uses the difference in time of arrival (ToA) of the RF signal from multiple transmitters at known positions. Similarly, PAL [11] uses ToA between the UWB signals to a plurality of receivers to determine the position. Cricket [12] and AHLoS [13] utilize propagation delays between RF and ultrasound signals to estimate the location of wireless devices. These solutions usually require tight synchronization of time and equipment extra ranging hardware, which limits their applicability.

Angle-of-arrival (AoA) based techniques make use of multiple antennas to estimate the angle at which the signal is received, and their geometric relationship to locate the wireless transmitter. Even with access to the information of the raw signal using 8 antennas [14], they can be derailed in rich multipath indoor environments where the direct signal path is often blocked. To solve this problem, some techniques based AoA require 6–7 sophisticated antenna systems [15], not possible in low resolution WiFi hardware products. Recently the Array-Track system [16] achieves a very precise localization using the AoA calculated from a rectangular array of 16 antennas. Arraytrack uses spatial and temporal smoothing, and time-based AoA grouping to suppress the effect of multipath on the location estimation. Although these techniques will certainly improve the accuracy of localization, Arraytrack stops to identify the actual angle of the direct path, particularly within the constraints of linear antenna arrays offered by the hardware WiFi products. SpinLoc [17] and Borealis [18] require the user to do a complete a $360° \pm$ turn.

2.2.2   Fingerprinting-Based Indoor Localization

Fingerprinting based solutions are established upon a general paradigm that, for each state instance $p \in P$, there is a unique corresponded "fingerprint" vector $v_p \in V$.

When this v_p is detected again at another time, we know the system is currently at state instance p. When this paradigm is specialized in the localization problem, this is fingerprint-based localization.

In outdoor environments, most wireless signals are propagated in Line-of-Sight (LoS) mode. It provides well suitable environment for model-based localization, i.e. precise timing or ranging.

However, in indoor environments, the multipath propagation has the major effect. The received signal at the endpoint is the sum of a larger number of reflected signals. It is difficult to separate these reflections, and hence hard to adopt model-based system.

Fingerprint-based schemes shortcut this difficulty by directly focusing on the signal distribution at specific positions.

Although it requires nontrivial human labor to site survey the signal map, its environmental-robustness still demonstrates its strong practicability to both research and industry communities.

In the early development stage of fingerprint based localization, there are few assistant techniques to improve the accuracy. With the rapid spreading of smart phone, more hybrid solutions appear.

a. Fingerprint Types

Among various approaches, RADAR [19] is the most famous and influential Wi-Fi fingerprint based localization system. It models the localization as a min-distance problem. In the deployment phase, the indoor environment is sampled by 2×2 m grid. For each grid, the RSS values from multiple APs are collected as the fingerprint for this grid. In the localization phase, the candidate position is the one which has the smallest "vector distance" to the fingerprint.

Since the human body may block the signal, the fingerprint shows large variance for different directions at the same spot. RADAR handles this problem by requiring sampling each position in 4 directions to alleviate this problem. The RADAR gains great success. It could achieve an average accuracy within 2 m, and most of fingerprint-based localization could be seen as variants of RADAR.

Horus [20] is widely accepted to be the most accurate Wi-Fi based fingerprint localization system. Its high accuracy relies on additionally deployed Wi-Fi sniffer devices. The additional Wi-Fi sniffers help create find-grained local RSS propagation model, which significantly improves the fingerprint accuracy. In experimental environment, the accuracy could be within 1 m.

In wireless signal domain, FM radio signal may be a promising medium for location fingerprint. The work proposed in [21] shows a FM radio signal based indoor localization system. The main protocol follows the RADAR scheme. The evaluations show that the FM radio RSS has much higher location distinction ability. The authors in [22] proposed an automatic FM fingerprint system by interpolating the RSS distribution map according to known indoor floor plan and pre-defined propagation model.

Wireless signal is not all about RSS. Channel State Information (CSI) or Channel Frequency Response (CFR) is another aspect of the signal. It is more find-grained PHY layer information than RSS. It reveals the amplitude and phase of wideband subcarriers. The works proposed in [23] and [24] proposed channel impulse response fingerprint based indoor localization. PinLoc [17] presents the first CSI-based indoor localization in rich multipath environment. These methods can generally achieve sub-meter fingerprint accuracy. However, since the channel response change very quick along spatial displacement, the fingerprints should be densely sampled.

Magnetic field also could be a type of fingerprint. [25] demonstrates a geomagnetism deviation based indoor localization system. The system is based on a discovery that the massive use of steel frame in modern construction alters the geomagnetism direction within the building. By recording the geomagnetism deviation for each spot, the geomagnetism could be used as the fingerprint. To measure the geomagnetism deviation, the authors built a rotational magnetometer. The evaluation shows strong de-correlation between spots.

Sound spectrum could be a natural fingerprint for indoor rooms. [26] proposed a room-level localization solely base on acoustic spectrum fingerprint. The advantage of acoustic fingerprint is that it does not require deploying infrastructures and it avoid the impact from the device variance problem commonly seen in Wi-Fi fingerprint solutions.

Besides the wireless signal, the ambient environment contains rich information to distinguish location. SurroundSense [27] is such a system. It traits the ambient sound spectrum, background color, and the like, as the "ambient fingerprint" to distinguish those spatial-nearby yet contextually-distant positions. A demonstrative example could be determine whether the user is at a bar or cafe, which are nearby according to GPS signal. Unloc [28] develops the idea of SurroundSense. It unified these ambient fingerprints as "Organic Landmarks" including sound spectrum, luminance, magnetic deviation, Wi-Fi landmarks, etc.

The localization for passive RFID tags is also an important problem. PinIt [29] proposed a RFID passive tag localization solution by densely deploying reference tags. PinIt uses synthetic aperture radar (SAR) technique to acquire the multipath profile for each reference tag. When the test tag is near a reference tag, their multipath profiles will be similar. By finding the reference tag which has the smallest difference, the position of test tag is determined.

b. *Crowdsourcing*

Crowdsourcing is a popular strategy trend that a difficult task is transformed and decomposed to many small and easy tasks. Each participant solves an easy task. All their efforts are then grouped together to have the difficult one finished. For fingerprint based indoor localization, the most difficult part is the tedious site survey procedure. If we could crowdsourcing this task, the fingerprint-based localization would become easy.

LiFS [30] proposed a system which employs the crowdsourcing idea. The major insight behind LiFS is that the adjacent fingerprint graph is highly similar to the floor plan routing graph. LiFS proposed an algorithm to automatically match these two graphs. In this way, it could eliminate the specific site survey task. Users only need to install and open an App in their smartphone. When people are moving in the environment, the App will collect the RSS values along the walking and send them to the localization server. When adequate amount of data is collected, the fingerprint map is generated.

SENIL [31] proposed a passive crowdsensing based system to automatically generate the radiomap. It uses graph matching algorithm to match the fingerprint map and floor plan. SENIL is deployed in the AP end. When there is communication traffic, the client is simultaneously contributing to the fingerprint collection task. Therefore, it does not require the client to install specific App, which significantly improves the crowdsourcing contribution rate.

c. *Hybrid Solutions*

The work proposed in [32] is a hybrid indoor localization system combining the Wi-Fi fingerprint and acoustic ranging. The acoustic ranging accuracy is much higher than Wi-Fi fingerprint. By combining both the coarse-grained Wi-Fi fingerprint localization and short range accurate acoustic ranging into one single optimization framework, it did push the limit of Wi-Fi localization accuracy.

Zee [33] is a combination of RADAR and inertial navigation based indoor localization. It proposed a hybrid localization framework, which supports various indoor localization schemes to work with inertial navigation. Since the position transition in indoor environments is restricted by the indoor structure, Zee proposed a Neuron-Network based user position interference scheme, which can correct the error accumulation of inertial navigation.

Walkie-Markie [34] proposed a hybrid solution based on inertial navigation and a spatially robust Wi-Fi landmark, called Wi-Fi Marks. Rather than stationary Wi-Fi fingerprint for each point, Wi-Fi Marks only exists at special position, where the RSS for specific AP reaches its local peak. These Wi-Fi Marks are sparsely distributed in the environment, and they serves as anchor points. The current position of user is obtained by joining the initial trajectory and the absolute position of anchors.

d. *Passive Fingerprint*

LANDMARC [35] is a device-free localization scheme. It built upon an active RFID tag array. When the tracking object is moving within the detection field, the object will cause signal strength variance around it. Reflecting on the nearby tags, the RSS value will show observable anomaly, which indicates the object is nearby.

It is currently rare to use the Fingerprint approach for multiple-objects scenario. SCPL [36] proposed an fingerprint based crowd counting system. SCPL requires deploying a large amount of transceivers around the room. The people are detected and counted by detecting the deviations between transceiver-transceiver links.

Fingerprint can be more general. It could be used to passively identify environmental events. Wi-Vi [37] proposed a through-wall people behavior recognition system. Sophisticated techniques are employed to remove the strong reflection from the wall. After that, Wi-Vi can detect the moving object (or people). By identifying the specific moving pattern, which is also fingerprint, the people outside the wall can know the current status inside the wall.

Wi-See [38] proposed a passive gesture recognition system feasible in whole-home area. It employs similar techniques to focus only on moving object, and the distinction resolution can be as small as an arm. Therefore, it could be used to sense gestures. By a simple training procedure, it can passively recognize the predefined gestures.

e. *Rethinking the Fingerprint*

Fingerprint based solutions gain huge development during past decade, however, it is perfect. There are various trade-offs to compromise it. Here we have a brief review on the trade-offs.

1. Deployment Cost v.s. Accuracy

The most challenging problem in fingerprint is that, it requires huge effort to do site survey. For an appropriate medium (such as Wi-Fi or FM radio), the denser the site survey, the higher accuracy. Compared to the most accurate Horus [20], the less accurate RADAR [19] is much more cost efficient and welcomed by both the research and industry communities. During the past decades, various methods have been proposed to reduce the site survey density without sacrificing accuracy. However, these methods are not generally available. For example, LANDMARC [35] may raise high cost for normal use, and LiFS [30] is not suitable for complex environments.

2. Fingerprint Medium v.s Accessibility

The fingerprint medium is another trade-off. Channel impulse response (CIR) or channel state information (CSI) based localization can provide high accuracy. However, this information is not accessible to commercial NIC. FM radio is relatively better than Wi-Fi signal, but it is only accessible to few mobile phones. Currently modern smartphones equip with magnetometers. The magnetometer based fingerprint requires a long sampling time. It is not convenient for common users.

3. Accuracy v.s Security

Fingerprint based localization has severe security problems. An attacker can forge an attacking beacon with the same MAC address or identity. When the victim detects this false beacon, the algorithm will provide wrong locations. To overcome this security risk, some methods [39] use channel throughput or transmission success rate as the fingerprint. In these schemes, the localization time is much longer and the accuracy is degraded.

4. Fingerprint v.s Model

The main advantage of the fingerprint-based solution over the model-based solution is that it bypasses the complex propagation modeling procedure in large complex area.

However, in a local small area, the propagation model could be easily established, while the fingerprint-based does not leverage this information.

There are some state-of-art approaches [22] trying to combine the local propagation model with large-scale fingerprints. These combinations are still conservative, contributing few to the accuracy.

References

1. Wangsness RK (1986) Electromagnetic Fields. Wiley-VCH
2. Zhu X, Li Q, Chen G, APT: Accurate Outdoor Pedestrian Tracking with Smartphones. In: IEEE INFOCOM, 2013.
3. Lim H, Kung LC, C. Hou J, Luo H (2010) Zero-Configuration Indoor Localization over IEEE 802.11 Wireless Infrastructure. ACM Wireless Networks 16:405–420
4. Madigan D, Elnahrawy E, Martin RP, Ju W-H, P. Krishnan, A.S. Krishnakumar, Bayesian Indoor Positioning Systems. In: IEEE INFOCOM, 2005. pp 1217–1227
5. Chintalapudi K, Iyer AP, Padmanabhan VN, Indoor Localization Without the Pain. In: ACM MobiCom, 2010. pp 173–184
6. Wu K, Xiao J, Yi Y, Gao M, Ni L, Fila: Fine-grained Indoor Localization. In: IEEE INFOCOM, 2012.
7. Nandakumar R, Chintalapudi KK, Padmanabhan V, Centaur: Locating Devices In an Office environment. In: ACM MobiCom, 2012.
8. Youssef M, Youssef A, Rieger C, Shankar U, Agrawala A, Pinpoint: An Asynchronous Time-based Location Determination System. In: ACM Mobisys, 2006.
9. Werb J, Lanzi C (1998) Designing A Positioning System for Finding Things and People Indoors. IEEE Spectrum.

10. Liu F, Cheng X, Hua D, Chan D, TPS: A Time-based Positioning Scheme for Outdoor Sensor Networks. In: IEEE INFOCOM, 2004.
11. Fontana RJ, Richley E, Barney J, Commercialization of An Ultra Wideband Precision Asset Location System. In: ICUWB, 2013.
12. Priyantha NB The Cricket Indoor Location System.
13. Savvides A, Han C-C, B M Dynamic Fine-grained Localization in Ad-hoc Networks of Sensors. In: ACM MobiCom, 2001.
14. Xiong J, Jamieson K, SecureAngle: Improving Wireless Security Using Angle-of-arrival Information. In: HotNets, 2010.
15. Niculescu D, Nath B, VOR Base Stations for Indoor 802.11 Positioning. In: ACM MobiCom, 2004.
16. Xiong J, Jamieson K, Arraytrack: A Fine-grained Indoor Location System. In: USENIX NSDI, 2013.
17. Sen S, Choudhury RR, Nelakuditi S, Spin Once to Know Your Location. In: Hotmobile, 2012.
18. Zhang Z, Zhou X, Zhang W, Zhang Y, Wang G, Zha BY, Zheng H, I Am the Antenna: Accurate Outdoor AP Location Using Smartphones. In: ACM MobiCom, 2011.
19. Bahl P, Padmanabhan V, RADAR: An In-building RF-based User Location and Tracking System. In: IEEE INFOCOM, 2000. pp 775–784
20. Youssef M, Agrawala A (2008) The Horus Location Determination System. ACM Wireless Networks 14:357–374
21. Chen Y, Lymberopoulos D, Liu J, Priyantha B, FM-based Indoor Localization. In: ACM Mobisys, 2012. pp 169–182
22. Yoon S, Lee K, Rhee I, FM-based Indoor Localization Via Automatic FingerprintDB Construction and Matching. In: ACM Mobile, 2013. pp 207–220
23. Nezafat M, Kaveh M, Tsuji H (2006) Indoor Localization Using a Spatial Channel Signature Database. Antennas and Wireless Propagation Letters. IEEE 5:406–409
24. Jin Y, Soh W-S, Wong W-C (2010) Indoor localization with Channel Impulse Response Based Fingerprintand Nonparametric Regression. Wireless Communications. IEEE Transactions on 9:1120–1127
25. Chung J, Donahoe M, Schmandt C, Kim I-J, Razavai P, Wiseman M, Indoor Location Sensing Using Geo-magnetism. In: ACM MobiSys, 2011. pp 141–154
26. Tarzia SP, Dinda PA, Dick RP, Memik G, Indoor Localization without Infrastructure Using the Acoustic Background Spectrum. In: Proceedings of the 9th international conference on Mobile systems, applications, and services, 2011. pp 155–168
27. Azizyan M, Choudhury RR (2009) SurroundSense: Mobile Phone Localization Using Ambient Sound and Light. ACM SIGMOBILE Mobile Computing and Communications Review 13:69–72
28. Wang H, Sen S, Elgohary A, Farid M, Youssef M, Choudhury RR, No Need to War-drive: Unsupervised Indoor Localization. In: Proceedings of the 10th international conference on Mobile systems, applications, and services, 2012. pp 197–210
29. Wang J, Katabi D, Dude, Where's My Card?: RFID Positioning that Works with Multipath and Non-line of Sight. In: SIGCOMM, 2013. pp 51–62
30. Yang Z, Wu C, Liu Y, Locating in FingerprintSpace: Wireless Indoor Localization with Little Human Intervention. In: ACM MobiCom, 2012. pp 269–280
31. Jiang Z, Zhao J, Han J, Tang S, Zhao J, Xi W, Wi-Fi FingerprintBased Indoor Localization without Indoor Space Measurement. In: ACM MobiCom, 2013. pp 384–392
32. Liu H, Gan Y, Yang J, Sidhom S, Wang Y, Chen Y, Ye F, Push the Limit of Wifi Based Localization for Smartphones. In: Proceedings of the 18th annual international conference on Mobile computing and networking, 2012. pp 305–316
33. Rai A, Chintalapudi KK, Padmanabhan VN, Sen R, Zee: Zero-effort Crowdsourcing For Indoor Localization. In: Proceedings of the 18th annual international conference on Mobile computing and networking, 2012. pp 293–304

34. Shen G, Chen Z, Zhang P, Moscibroda T, Zhang Y, Walkie-Markie: Indoor Pathway Mapping Made Easy. In: USENIX NSDI 2013. pp 85–98
35. Ni LM, Liu Y, Lau YC, Patil AP (2004) LANDMARC: Indoor Location Sensing Using Active RFID. ACM Wireless Networks 10:701–710
36. Xu C, Firner B, Moore RS, Zhang Y, Trappe W, Howard R, Zhang F, An N, Scpl: Indoor Device-Free Multi-subject Counting and Localization Using Radio Signal Strength. In: ACM IPSN, 2013. pp 79–90
37. Adib F, Katabi D, See Through Walls with WiFi. In: Proceedings of the ACM SIGCOMM 2013 conference on SIGCOMM, 2013. pp 75–86
38. Pu Q, Gupta S, Gollakota S, Patel S, Whole-home Gesture Recognition Using Wireless Signals. In: Proceedings of the 19th annual international conference on Mobile computing \& networking, 2013. pp 27–38
39. Meng W, Xiao W, Ni W, Xie L, Secure and Robust Wi-Fi Fingerprinting Indoor Localization. In: IEEE IPIN, 2011. pp 1–7

Chapter 3
Critical State and Twins

3.1 Mutual Interference Between Tags

The interference between tags, known as a challenging issue when achieving high access rate on tags, has gained attentions. Steven M. Weigand and Daniel M. Dobkin [1] conducted extensive experiments on the interference between tags. Their experiment scenario is shown in Fig. 3.1. They deployed two tag arrays in two planes. In the front plane (near the reader), the tags are deployed in a 9×3 array. In the back plane, the tags are deployed in a 6×3 array. The columns were spaced 30 cm apart and the rows 5 cm apart. They used a WJ Communications MPR6000 PC-card reader for interrogation (operating at 924 MHz and with a nominal output power of 27 dBm), the reader is connected with a linearly-polarized 9-dBi patch antenna, which was placed 50–60 cm from the front plane. Such a deployment is a simulation on densely packed items, whose attached tags are close to each other. They investigated the successful reading times for each tag. The result reported in Table 3.1 show that the appearance of back tag array incur a negative effect on front tag array in terms of reading rate. Furthermore, they found that when varying the interplane gap, the resulting tag read shows a similar change on the two tag arrays. The front plane is strengthened in a periodicity of roughly 16 cm, corresponding to about 1/2 wavelength at the operating frequency.

They attempted to theoretically analyze this phenomenon. They propose to use an ideal-dipole-like model to represent the tag antenna, which is a greatly simplified antenna element to simulate the short, capacitive-loaded antennas often used for RFID tags. We show their model in Fig. 3.2. Note that the structure of tag in their model is simplified into two short bars symmetrically placed on the two sides of the tag chip. With this model, Steven M. Weigand and Daniel M. Dobkin show the feasibility of periodical variation on the electromagnetic field when two tag arrays are approaching. Later, some researchers also model the mutual interference between tags over other practical scenarios. For example, T. Ye et al. modeled a specific application scenario for stacked tags. The authors also demonstrated a significant mutual interference between tags when they are packed closer to each other. They presented a theoretical model, in which multiple tag antennas were stacked

© The Author(s) 2014
J. Han et al., *Device-Free Object Tracking Using Passive Tags,* SpringerBriefs
in Electrical and Computer Engineering, DOI 10.1007/978-3-319-12646-3_3

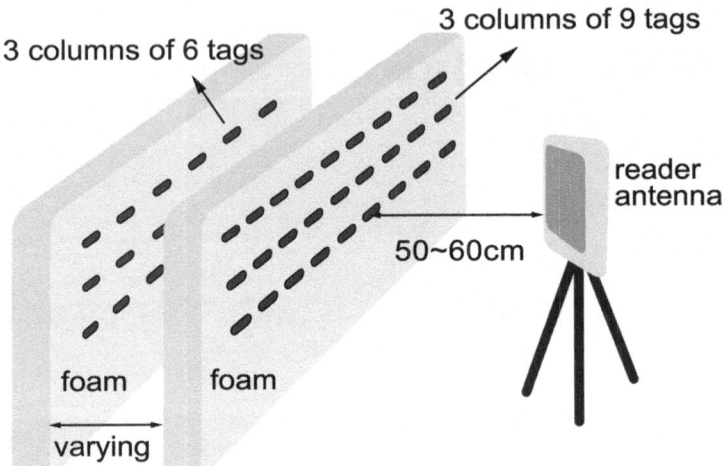

Fig. 3.1 Experiment configuration in [1].

vertically with a certain distance in between, as shown in Fig. 3.3. In this model, each tag is replaced by an identical circular loop. They analyzed the mutual coupling between two adjacent loops and then extended to the multiple stacked loops. Utilizing this model, the authors can explore the tags that are unable to read due to the mutual coupling, namely *weak spots* in the stack. The experiment and emulation results fit with that derived from the theoretical analysis.

Nevertheless, prior approaches place the tags in specific patterns and model the tag in an ideal shape. In certain cases, such approximations can retrieve the satisfied fitting results, but do not help to explicitly reflect the RF and antenna performance of real implementations. Based on our observation, the mutual interference between tags take obvious effect in the near field. In this section, we setup a more general model to formulate the mutual interference. This model is suitable for the dense deployment, in which the inductive coupling dominates the mutual interference. We plot the model in Fig. 3.4.

In this model, we adopt the tag approximation in [2], where the tag antenna is represented as an equivalent circular loop. Such an approximation is commonly used by prior works and the circular loop does not involve any specific structure of real tags. Our model is more general than previous models by placing two nearby tags, i.e., two circular loops with certain distance and angel between their center positions. Without loss of generality, we set Tag1 at the point (0, 0, 0) and Tag2 at

Table 3.1 Tag read v.s. interplane gap

Interplane gap (cm) tag read	8 (¼ λ)	12	16 (½ λ)	20	24 (¾ λ)	28	32 (λ)	36
Front plane	18	7	8	21	22	12	12	20
Back plane	0	0	2	4	2	0	0	0

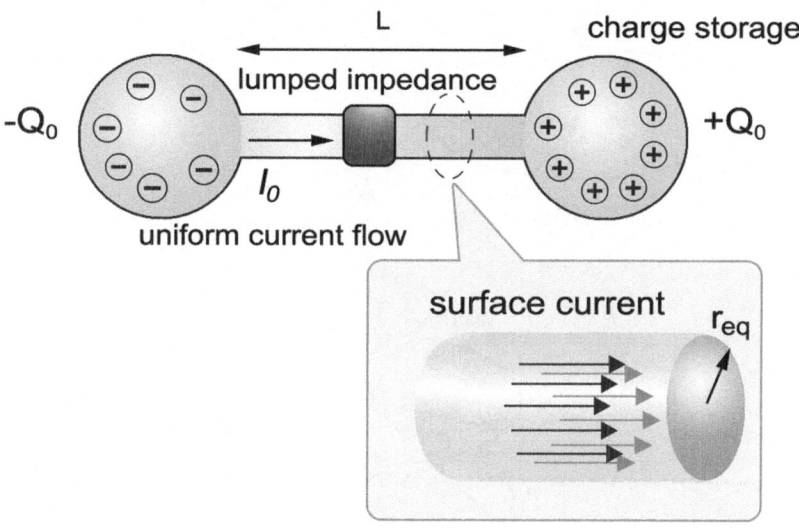

Fig. 3.2 Ideal-dipole-like model of tag antenna in [1].

(0, T$\sin\alpha$, T$\cos\alpha$), where the $\vec{T}$ is the vector from the center of Tag1 to that of Tag2 and the α is the angle between Z axis and $\vec{T}$.

We denote the current on the loop of Tag1 as I_1. According to the Biot-Savart Law [3], at each point of the loop of Tag1, the steady I_1 can induce a magnetic field on Tag2. Suppose dB_2 is the magnetic induction induced by I_1 at the point P of Tag1 loop, it can be calculated as

$$dB_2 = \frac{\mu_0 I_1 \overline{dl} \times (\vec{T} - \vec{R})}{4\pi \left| \vec{T} - \vec{R} \right|^3} \tag{3.1}$$

Here, the μ_0 is the magnetic permeability of vacuum, $\overline{dl}$ is a vector whose magnitude is the length of the differential element of the wire in the direction of conventional current, $\vec{R}$ is the radius vector along the angle θ, and $\vec{T} - \vec{R}$ is the positional

Fig. 3.3 The model of stacked tags

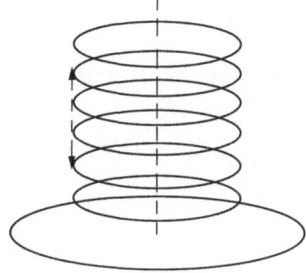

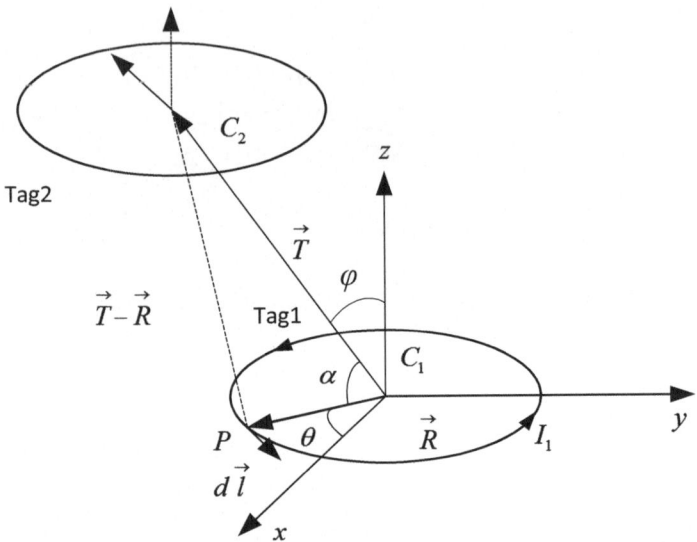

Fig. 3.4 Mutual interference model for two passive tags

vector starting from the current segment on the Tag1's antenna loop and pointing to the center of Tag2's loop. Since the induced magnetic induction can be overlapped, we can calculate the entire magnetic induction induced by I_1 as the integration of the induced magnetic induction along the loop of Tag1:

$$\overrightarrow{B_2} = \frac{\mu_0}{4\pi} \oint_0^{2\pi} \frac{I_1 \overrightarrow{dl} \times (\overrightarrow{T} - \overrightarrow{R})}{\left|\overrightarrow{T} - \overrightarrow{R}\right|^3} d\theta \tag{3.2}$$

Suppose that the element vectors of x, y, and z axis are $\overrightarrow{e_x}, \overrightarrow{e_y}, and \ \overrightarrow{e_z}$, respectively. The coordinate of P is $(Rcos\theta, Rsin\theta, 0)$. The coordinate of the center of Tag2 C_2 is $(0, Tsin\varphi, Tcos\varphi)$. Therefore, $\overrightarrow{dl} \ and \ \overrightarrow{T} - \overrightarrow{R}$, can also be represented as

$$\overrightarrow{dl} = dl(-\overrightarrow{e_x}sin\theta + \overrightarrow{e_y}cos\theta) = Rd\theta(-\overrightarrow{e_x}sin\theta + \overrightarrow{e_y}cos\theta)$$

$$\overrightarrow{T} - \overrightarrow{R} = -\overrightarrow{e_x}Rcos\theta + \overrightarrow{e_y}(Tsin\varphi - Rsin\theta) + \overrightarrow{e_z}Tcos\varphi \tag{3.3}$$

Replacing the $\overrightarrow{dl} \ and \ \overrightarrow{T} - \overrightarrow{R}$ in Eqn. 3.2 with those in Eqn. 3.3, we have

$$\overrightarrow{B_2} = \frac{\mu_0}{4\pi} \oint_0^{2\pi} \frac{I_1 R\left(\overrightarrow{e_x}Tcos\theta cos\varphi + \overrightarrow{e_y}Tsin\theta cos\varphi + \overrightarrow{e_z}(-Tsin\theta sin\varphi + R)\right)}{\left(R^2 + T^2 - 2RTsin\varphi sin\theta\right)^{3/2}} d\theta \tag{3.4}$$

Since only the magnetic induction vertical to the plane of loop2 can induce a current in Tag2, we consider the component of $\overrightarrow{B_2}$ along the z axis. After further simplifying Eqn. 3.4, we have

$$\overrightarrow{B_2} = \frac{\mu_0}{4\pi} \oint_0^{2\pi} \frac{I_1 R(-T sin\theta sin\varphi + R)}{\left(R^2 + T^2 - 2RT sin\varphi sin\theta\right)^{3/2}} d\theta \tag{3.5}$$

With the $\overrightarrow{B_2}$, we are able to compute the *mutual inductance M* between two tags. The mutual inductance is a measurement on the coupling effect between two loops in this model. We assume the magnetic flux induced by the current I_1 in Tag1 passing through Tag2's loop as $\Psi_{21}(I_1)$. According to the magnetic flux definition, M_{21} can be written as

$$M_{21} = \frac{\Psi_{21}(I_1)}{I_1} = \frac{\overrightarrow{B_2} A_2}{I_1} \tag{3.6}$$

where A_2 represents the area of the Tag2's loop surface where the magnetic lines of flux pass through. Merging Eqns. 3.5 and 3.6, we have

$$M_{21} = \frac{\mu_0 A_2}{4\pi} \oint_0^{2\pi} \frac{R(-T sin\theta sin\varphi + R)}{\left(R^2 + T^2 - 2RT sin\varphi sin\theta\right)^{3/2}} d\theta \tag{3.7}$$

Following the Lenz law [3], the current I_{T_2} in Tag 2 induced by I_1 is also determined by the equivalent RCL circuit of Tag2. The direction of I_{T_2} is opposed to that of I_1:

$$I_{T_2} = \frac{-j\omega M_{21} I_1}{R_{chip} + 1/j\omega C + j\omega L} = -bM_{21} I_1 \tag{3.8}$$

where R_{chip}, C, L are the chip impedance, capacitance, and inductive impedance of Tag 1, respectively, the b denotes the hardware–relative part of above equation.

Assume that I_{01} and I_{02} are the currents generated by harvesting the RF signals from the reader in Tag1 and Tag2. The current in Tag 2 can be represented as $I_2 = I_{02} - bM_{21}I_1$. While the current in Tag2 can also induce a current in Tag1, and hence incur a mutual inductance M_{12}. According to the *Reciprocity Theorem* combining the Ampere's law and *Biot-Savart* law [3], the two mutual inductances are equal, i.e., $M = M_{12} = M_{21}$. Thus, $I_1 = I_{01} - bM_{12}I_2$.

In our model, the range of near field is very short, e.g. several millimeters between two tags. On the other hand, the distance from the reader's antenna to the tags is much larger than that between two tags. From the viewpoint of the reader, the two tags are in a same distance from the reader's antenna. Thus, the currents I_{01} and I_{02} are equal and with the same direction, i.e. $I_{01} = I_{02}$. Combining with the equivalence between M_{12} and M_{21}, this means for the two tags, $I_1 = I_2$.

Obviously, if following this "structure-oblivious" model, the mutual interference will have equivalent effects on two tags. That is, the current in both tags' antennas should be reduced with the same amount. Therefore, when two tags are approaching with each other, they should become unreadable almost simultaneously at a certain distance. However, we observe a conflict in our real experiment, which will be elaborated in Sect. 3.2.

3.2   Critical State and Twins

The mutual interference between two tags inspires us to produce a new motion detector. If we place two tags parallel with each other and leave a short distance in between, as aforementioned in Sect. 3.1, the two tags may become unreadable in a certain distance due to the mutual interference. We define the state that the two tags JUST become unreadable as *critical state*. In practice, we find that it is also possible that only one tag becomes unreadable while another does not. We will explain the latter phenomenon in this section.

As we know, the passive tag harvests the required energy from the RF signals emitted from the reader. If the energy accumulated from the harvesting is counteracted by the mutual interference, the passive tag cannot operate and modulate. As a result, the reader cannot identify the tag.

This phenomenon can be utilized to achieve device-free motion detection. We deliberately create a critical state for two close tags, and deploy the pair of tags in the area of interests. If an object or human being moves around the pair of tags, it will reflect or refract some RF signals to the tag. The signals casted into the tag's antenna will provide a certain amount of energy to the tag. Since the tag is in the critical state, a very small portion of energy may be sufficient for one tag or two tags to operate and become readable again. Such a state transfer can be easily captured by the reader, indicating a motion happening around the pair of tags. Such a state transfer is termed as *state jumping*. By observing the state jumping and its amount, moving objects or human beings can be detected. We name such a pair of tags as *Twins* or a Twins pair hereafter (Fig. 3.5).

However, when we verify the feasibility of Twins in motion detections, we find that in the critical state of Twins, it is also possible that only one tag becomes unreadable and another tag can be identified normally. We use 20 randomly selected off-the-shelf tags modeled Impinj E41-b in our preliminary experiments. The tags are denoted as A_1, A_2, ..., A_{10}, B_1, B_2, ..., B_{10}. We form ten Twins (A_1, B_1), ..., (A_{10}, B_{10}). For each Twins, we vary the transmission power of reader and record the minimum power needed for reading the tags and report the result in Fig. 3.6. In these experiments, we fix the distance between two tags to 10 mm and the distance between tags and the reader's antenna to 2 m.

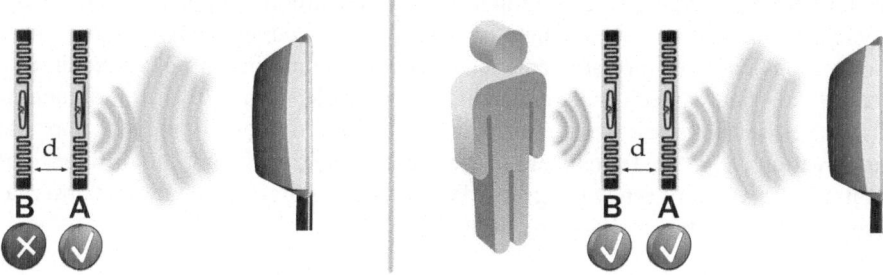

Fig. 3.5 Critical state of twins

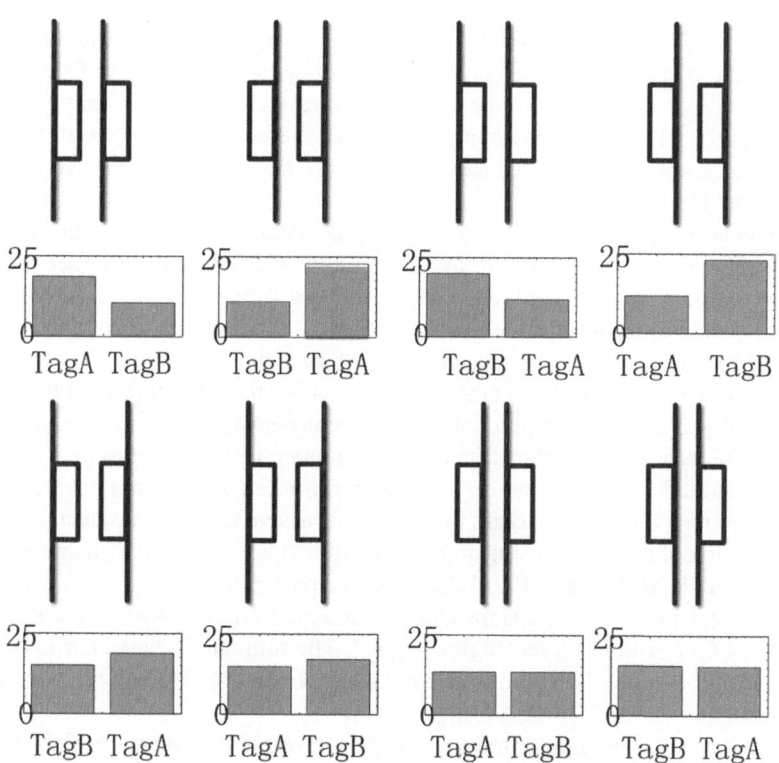

Fig. 3.6 The layout of Twins and the minimum P_{TX} used for reading them

Increasing the transmission power for the reader yields an increased current at the tag. We can use the value of minimum power to scale the current induced in the tag for operation. In Fig. 3.6, the geometry shape of tags is simplified. We define the tag with its IC as the closest part to another tag than its other parts as the *Rear-tag* in Twins, like Tag A of the first Twins in the first row. Correspondingly, Tag B will be the *Fore*-tag in that Twins. We report the average minimum powers for each tag. There is an interesting phenomenon. In certain types of deployed Twins pairs where the two tags are with the same antenna direction, the minimum power required to break the critical state is quite different, e.g. about 10 dbm in this experiment, as illustrated the Twins pairs in the first row of Fig. 3.6. For those Twins with opposite tag directions, e.g. the Twins pairs in the second row of Fig. 3.6, the difference of minimum power between two tags in a Twins is negligible. Obviously, this observation is opposed to the analysis on our model.

3.3   T-match Structure and Structure-Aware Model

Currently, most passive tags in the market use a half-wave dipole antenna, for example the typical commercial passive tag, Impinj E-41b shown in Fig. 3.5. The length of antenna is related to the wave length of the reader. It is known that $\lambda/2$, i.e. 16 cm for 915 MHz, allows the tag to have the best performance in backscattering communications. To reduce the physical size, manufactures usually use a *meandered dipole* antenna.

However, simply using the meandered dipole antenna may cause a mismatching between the dipole and the IC of tag. A common solution is to introduce a shorter antenna to balance the capacitive IC load, forming a T-match structure. As illustrated in Fig. 3.5, the impedance of the longer meandered dipole (with the length of L) can be tuned by the introduced shorter dipole (with the length of a).

With the T-match structure in mind, we reconsider the mutual interference model between two tags. We attempt to adopt a "structure-aware" interference model that can provide reasonable explanation to the phenomenon of one-tag state jumping.

As illustrated in Fig. 3.7, we do not simply formulate a tag as a loop. Instead, we use a simplified T-match structure to model the passive tag with dipole antennas. In detail, we model the tag by using an electric dipole like a line and a magnetic dipole like a rectangle, as shown in Fig. 3.7a. Thus, the pair of tags with the same direction in a Twins can be remodeled as the example in Fig. 3.7b. In this model, a Twins has two lines (L_1, L_2) and two rectangles (S_1, S_2). The ultimate induced currents in the Twins are determined by these four conductors. Note that Tag1 is the Rear-tag in the Twins.

We assume that the reader induce coupling currents $I_{L_{01}}$ and $I_{L_{02}}$ in the two lines L_1 and L_2, respectively. Since the IC is connected to the magnetic dipole, the current in S_1 is approximately equal to the current activating the tag, i.e. $I_1 = I_{S_1}$ and $I_2 = I_{S_2}$. We have the following theorem.

Theorem: Given two tags *Tag1* and *Tag2* are placed as in Fig. 3.9b, let I_{S_1} and I_{S_2} be the current on S_1 and S_2, respectively. $I_{S_1} < I_{S_2}$

Proof: Since the distance from the line to the reader's antenna is almost the same, we have $I_{L_{01}} = I_{L_{02}} = I_{L_0} e^{j\omega t}$ where I_{L_0} is the complex amplitude. The area of the rectangle is $a \times b$. We use r to denote the distance from the rectangle to the line.

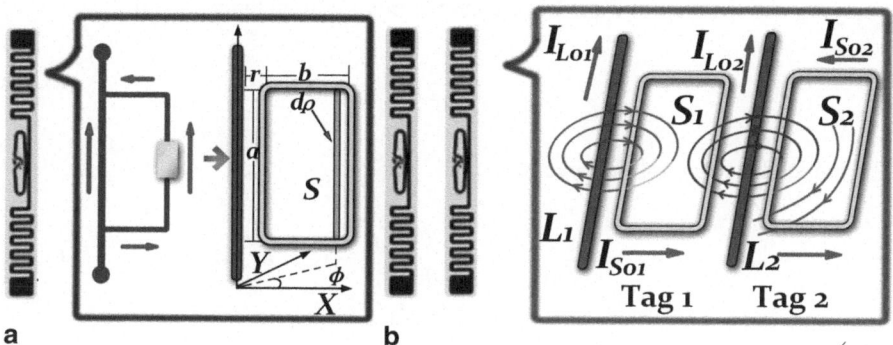

Fig. 3.7 Structure-aware modeling for the twins

Without loss of generality, we assume that $I_{L_{01}}$ and $I_{L_{02}}$ are in the same direction as Z axis. We then analyze the mutual interference and the induced currents in the tags.

a. Mutual Inductance in S_1

First, the inductance within S_1 is an integration of the mutual inductances between L_1 and S_1, between L_2 and S_1, and between S_1 and S_2.

1. Mutual inductance between L_1 and S_1

The current I_{01} will generate concentric circles around L_1, representing a magnetic induction B. For a point (ρ, φ, z) within the rectangle S_1, we have the magnetic induction B in this position using the *Biot-Savart* Law [3],

$$\vec{B} = \vec{e}_\varphi \frac{\mu_0 I_{L_{01}}}{2\pi\rho} \tag{3.9}$$

where μ_0 is the magnetic constant and $\vec{e}_\varphi$ is the unit direction vector of the magnetic induction.

We then get the entire magnetic flux caused by $I_{L_{01}}$ by calculating the integral of $\vec{B}$ within the rectangle S_1,

$$\Psi_{11} = \int_{S_1} \vec{B}_{S_1} d\vec{S} \tag{3.10}$$

For easy computation, we use $d\rho$ to denote the tiny rectangle facet primitive within the S_1. Suppose the a and b are the width and length of the rectangle, respectively. Therefore, $\vec{S} = ad\rho$. We calculate the integral of $\vec{B}$ from r to $r+b$

$$\Psi_{11} = \frac{\mu_0 I_{L_{01}} a}{2\pi} \int_r^{r+b} \frac{1}{\rho} d\rho = \frac{\mu_0 I_{L_{01}} a}{2\pi} \ln\left(\frac{r+b}{r}\right) \tag{3.11}$$

Applying above result to Eqn. 3.6, we get the mutual inductance M_{11} of L_1 and S_1 as

$$M_{11} = \frac{\Psi_{11}}{I_{L_{01}}} = \frac{\mu_0 a}{2\pi} \ln\left(\frac{r+b}{r}\right) \tag{3.12}$$

The induced electromotive force in S_1 is

$$\Delta E = -\frac{d\Psi_{11}}{dt} = -\frac{M_{11} dI_{L_{01}}}{dt} = -j\omega I_{L_0} e^{j\omega t} M_{11} \tag{3.13}$$

$$= -\frac{\mu_0 a j\omega I_{L_0} e^{j\omega t}}{2\pi} \ln\left(\frac{r+b}{r}\right)$$

Note that the direction of magnetic flux induced by L_1 is opposite to that of S_1. Thus, the induced electromotive force in S_1 should be negative, i.e. $E_{11} = -\Delta E$.

For simplicity, we assume that the equivalent resistance in S_1 and S_2 is the same, denoted as R. The current coupled by $I_{L_{01}}$ in S_1 is

$$I_{11} = \frac{E_{11}}{R} = \frac{\mu_0 a j\omega I_{L_0} e^{j\omega t}}{2\pi R} \ln\left(\frac{r+b}{r}\right) \tag{3.14}$$

2. Mutual inductance between L_2 and S_1

The coupling effect M_{21} between L_2 and S_1 can be derived in a similar way. The mutual inductance can be written as

$$M_{21} = \frac{\Psi_{21}}{I_{L_{01}}} = \frac{\mu_0 a}{2\pi} \ln\left(\frac{l+b}{l}\right) \tag{3.15}$$

However, the direction of magnetic flux of $I_{L_{02}}$ is same as that of S_1, augmenting the integrated magnetic flux. Similar to Eqn. 3.14, we have

$$I_{21} = -\frac{\mu_0 a j\omega I_{L_0} e^{j\omega t}}{2\pi R} \ln\left(\frac{l+b}{l}\right) \tag{3.16}$$

3. Mutual inductance between S_1 and S_2

According to *Reciprocity Theorems* [3], S_1 and S_2 is symmetric to each other. Each of them has an identical value of coupled current, namely $-I_H$, to the other, and the induced current is opposite the exciting current.

b. Inductance in S_2

Similar to Eqns. 3.14 and 3.16, we can calculate the coupled current I_{12} between L_1 and S_2, and the coupled current I_{22} between L_2 and S_2.

$$I_{12} = \frac{\mu_0 aj\omega I_{L_0} e^{j\omega t}}{2\pi R} \ln\left(\frac{2r+2b+l}{2r+b+l}\right) \tag{3.17}$$

$$I_{22} = \frac{\mu_0 aj\omega I_{L_0} e^{j\omega t}}{2\pi R} \ln\left(\frac{r+b}{r}\right) \tag{3.18}$$

where the l represents the distance between S_1 and L_2, and $l \approx d$.

c. Currents in S_1 and S_2

The currents in S_1 and S_2, denoted as $I_{S_{01}}$ and $I_{S_{02}}$ are induced by the RF signals from the reader's antenna. They can be calculated as:

$$I_{S_1} = I_{S_{01}} + I_{11} + I_{21} - I_H$$

$$= I_{S_{01}} - I_H + \frac{\mu_0 aj\omega I_{L_0} e^{j\omega t}}{2\pi R} \ln\left(\frac{r+b}{r}\right) - \frac{\mu_0 aj\omega I_{L_0} e^{j\omega t}}{2\pi R} \ln\left(\frac{l+b}{l}\right)$$

$$I_{S_2} = I_{S_{02}} + I_{12} + I_{22} - I_H$$

$$= I_{S_{02}} - I_H + \frac{\mu_0 aj\omega I_{L_0} e^{j\omega t}}{2\pi R} \ln\left(\frac{2r+2b+l}{2r+b+l}\right) + \frac{\mu_0 aj\omega I_{L_0} e^{j\omega t}}{2\pi R} \ln\left(\frac{r+b}{r}\right) \tag{3.19}$$

Note that S_1 and S_2 are in a same rectangle shape, which is a "structure-oblivious" model similar to the one we analyzed in Sect. 3.1. Thus, the value of coupled currents caused by S_1 and S_2 to another rectangle is identical, i.e., $I_{S_{02}} = I_{S_{02}}$. Consequently, $I_{S_2} > I_{S_1}$.

$I_{S_2} > I_{S_1}$ indicates that the rear tag cannot harvest sufficient energy due to the mutual effect from the front tag. This explicitly explain why the rear tag always becomes unreadable when a same-direction Twins is in a critical state. From above analysis, we also learn that the values of l and b are crucial to the generation of critical state. Since the l is the distance between S_1 and L_2, it is possible for us to leverage it to generate the critical state. When we adjust the l to the same scale as b, i.e. $l \approx b$, the difference between the I_{S_2} and I_{S_1} becomes distinct, such that triggering the critical state for the Twins is much easier. On the other hand, if $l \gg b$, i.e. the distance between two tags is sufficiently large, we can get

$$\lim_{\frac{b}{l}\to 0}\frac{l+b}{l} = 1, \lim_{\frac{b}{l}\to 0}\frac{2r+2b+l}{2r+b+l} = 1, \lim_{\frac{b}{l}\to 0}\ln\left(\frac{l+b}{l}\right) = 0$$

$$\lim_{\frac{b}{l}\to 0}\ln\left(\frac{2r+2b+l}{2r+b+l}\right) = 0, and\; I_{21} = I_{22} \tag{3.20}$$

In this case, $I_{S_1} \approx I_{S_2}$, resulting in nearly identical inductive coupling effect for the two tags to each other, and the critical state is hardly to be generated.

3.4   Twins Based Motion Detection

We perform experiments to validate the critical state and state jumping phenomenon. First, we form a Twins using two randomly selected tags A and B. Let l denote the distance between two tags, and D denote the distance from the reader to the Twins. There are two ways to generate the critical state in a controllable manner. First, we fix the distance l as 10 mm. The critical state of this Twins can be generated by tuning the transmission power of the reader, termed as critical power, P_{TX}. We use the UI of commercial RFID reader for this tuning. In detail, we scan the power spectrum from the maximum setting of P_{TX} from 32.5 to 10 dBm, until the critical state occurs in Twins. Second, we vary the value of l from 6 mm to 26 mm and try to yield a critical state for the Twins with a fixed D of 2 m. The experiment setup is illustrated in Fig. 3.8.

For the two tags used in our experiment, the critical state occurs when l<15 mm, as shown in Fig. 3.7. In the twins, Tag B is the Rear-tag. The experiment result shows that Tag B is always the one that is unreadable in the critical state. Correspondingly, Tag B shows a higher minimum value of P_{TX} than that of Tag A in Fig. 3.9. With reducing the value of l, the difference between the values of P_{TX} of the two tags is enlarged.

We invite a volunteer to move around the Twins staying in the critical state. We then plot the positions where the Twins detects the movement in Fig. 3.10. From this figure, we can geometrically sketch the detection range of Twins. It can be found that the effective detection range of Twins is relatively large, say nearly 2×1 m between the reader and Twins. Furthermore, we divide the experiment area into small cells, and count the number of state jumping events in each cell. We normalize and plot the counts in those cells in Fig. 3.11, which shows the sensitivity of Twins corresponding to its geometric distribution in the detection region. In a short summary, the effective detection range of Twins is about 1 m away from the Twins.

Fig. 3.8 Experiment setup

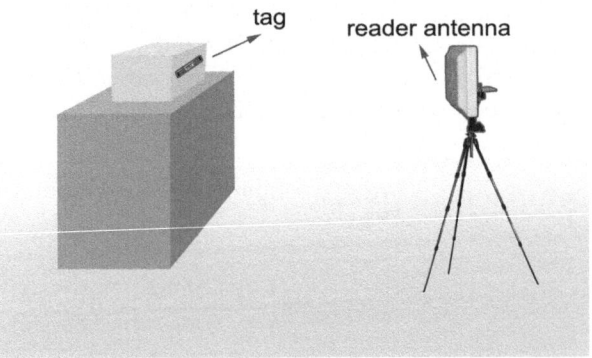

Fig. 3.9 Minimum P_{TX} vs. l

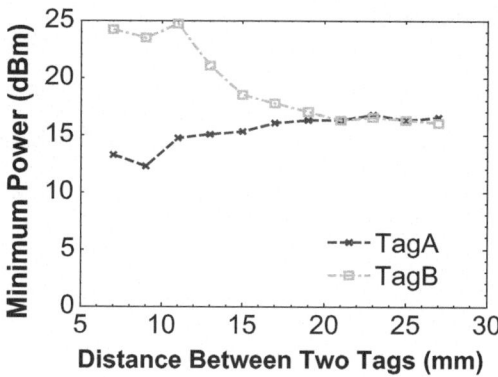

Fig. 3.10 # of state jumping

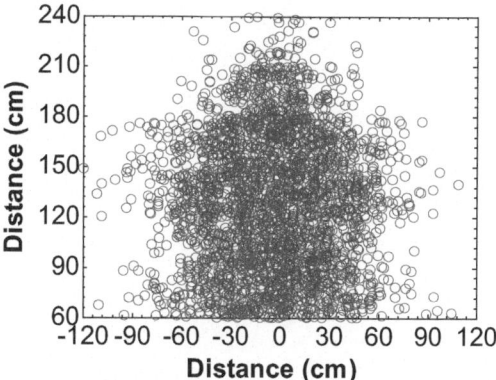

Fig. 3.11 Effective region

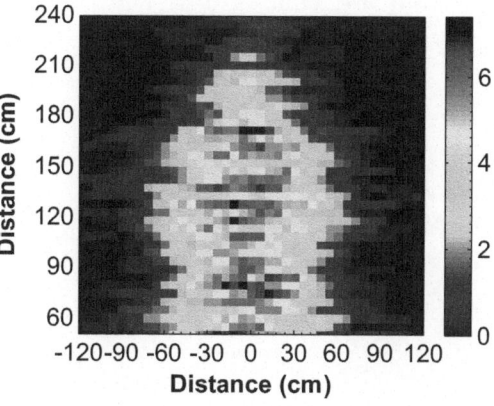

References

1. S.M. Weigand, D.M. Dobkin Multiple RFID Tag Plane Array Effects. In: Antennas and Propagation Society International Symposium, 2006. pp. 1027–1030
2. Chen X, Lu F, T.Ye T The 'Weak Spots' in Stacked UHF RFID Tags in NFC Applications. In: RFID, 2010.
3. Wangsness, RK (1986) Electromagnetic Fields. Wiley-VCH

Chapter 4
Dense Deployment Based Tracking System Using Twins

Utilizing Twins as the detector, we can reuse the deployed passive RFID systems for tracking an object. The basic idea is to leverage the critical state of Twins to capture the movement of objects or human beings. In this chapter, we propose a dense deployment based tracking scheme using Twins.

The tracking process comprises two phase, localization and tracking. In the first phase, the state jumping on Twins is used to pinpoint the location of moving object. In the second phase, we use a Kalman filter or particle filter based method to track the object. Currently, the Twins based method supports tracking single object only.

Suppose a number of Twins have been deployed in the given area, for instance, the pick aisles between two shelves with valuable items. We assume each Twins can be interrogated and identified by at least one reader. All the Twins are deployed in a grid pattern and the entire region is partitioned into cells, as shown in Fig. 4.1. We empirically set the distance between two Twins. The Twins grid forms a virtual graph $G=<V, E>$, where each cell is a vertex and two adjacent cells have an edge. If a state jumping is detected at a certain Twins pair, the corresponding cell will be marked in G.

4.1 Identifying Twins with a state jumping

As aforementioned, the moving object will trigger a state jumping on the Twins along its trace. The trace can be outlined if we can timely capture the state jumping incurred on those Twins pairs. The limit of current commercial passive RFID readers, however, raises a challenge to the detection. The design of existing RFID reader can only use a fixed transmission power for interrogation at any time point. So at any time, the reader triggers the critical state for only one single Twins pair. Since different Twins pair may have various critical power thresholds, the transmission power transfer from one Twins' power setting to another will be time-consuming. As a result, some state jumping will be missed and the trace cannot be successfully plotted. Thus, the ultimate objective of our tracking scheme is to design an efficient scheduling mechanism allowing the passive reader to timely react to the movement

© The Author(s) 2014

J. Han et al., *Device-Free Object Tracking Using Passive Tags*, SpringerBriefs in Electrical and Computer Engineering, DOI 10.1007/978-3-319-12646-3_4

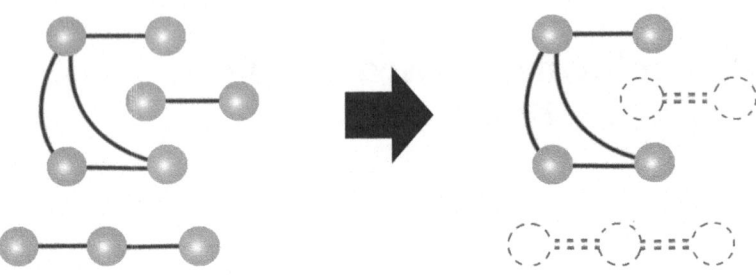

Fig. 4.1 Selecting G_s

of moving object and interrogate the Twins along its trace accordingly. In addition, we need to avoid the schedule starvation occurring on some Twins pairs if they have low priority in such scheduling.

We propose a polling algorithm, namely the Multi-level Priority Linked-List based (MPLL) algorithm. The basic idea of MPLL is to maintain multiple priority lists to indicate the sequence of the state jumping occurrence among Twins pairs. During the polling procedure, the Twins with more recent state jumping will be put into the list with higher priority. Assume that the system contains N Twins pairs. For each Twins, it has a tuple $t=<P_{TX, i}, T_i, P_i, S_i>$ in the database, where T_i denotes the sequence number of the Twins, $P_{TX, i}$ denotes the corresponding transmission power of reader's antenna to create the critical state of T_i. P_i and S_i are two bits representing the query priority and access status of T_i. In our algorithm, we adopt two-lists for the MPLL. The two lists are denoted as L_P and L_N. L_P is with higher priority than L_N, and contains the tuples of Twins pairs experiencing state jumping in last polling round. Here, a Twins pair will be inserted into in L_P if P_i is 1, or in L_N if $P_i=0$.

MPLL iteratively queries the Twins pairs in a Breadth-First Search (BFS) pattern in the two lists. Intuitively, MPLL starts the BFS query from the beginning of L_P list. Suppose that at a certain time, MPLL is querying a Twins pair T_i in L_P. MPLL checks whether it has experienced a state jumping in this round. If Yes MPLL further checks whether $P_i=1$, which means that this Twins pair is still in L_P. If the answer is still Yes, MPLL keeps the Twins pair in L_P, and then collects all its neighboring Twins pairs via G. For each neighboring Twins, MPLL performs the query as well. In this way, MPLL will interrogate as many Twins pairs nearby the moving object as possible. If the checked Twins does not record any state jumping, MPLL will move it to the low priority list L_N. Then MPLL moves to the successor of T_i. Above query is recursively executed within the two list.

MPLL may query some Twins multiple times due to the connection on G. To improve the efficiency, we design two query patterns for MPLL, *real query* and *virtual query*. We also introduce a state bit S_i to each tuple of Twins as a marker showing whether the Twins pair has been accessed in this polling round. The real query is only performed when the Twins pair is under the first query of current polling round. Before the execution, MPLL set $S_i=0$ for each T_i in the database. Once being queried by the reader, MPLL flips S_i to 1 for T_i. During the checking, if S_i is

0, MPLL launches a real query to T_i. If a state jumping incurs, MPLL sets $S_i=1$ and $P_i=1$ for T_i. On the other hand, $S_i=1$ indicates that the Twins T_i has been accessed by MPLL in this polling round. It is not necessary to perform a real query via the reader. MPLL will launch a virtual query and only check the value of P_i to judge whether T_i experiences a state jumping or not. Once MPLL finds that no state jumping incurs on T_i, T_i is moved from L_P to L_N.

When MPLL moves to the end of L_P, it starts the polling from the former terminating position at L_N. The polling in L_N does not need a BFS at each element. If the Twins has not experienced a state jumping, MPLL sets its $S_i=1$ and performs a real query. If this Twins reports a state jumping, MPLL sets its $P_i=1$ and moves the Twins to L_P. Otherwise MPLL directly moves to its successor in L_N to execute the polling, until it moves to the end of L_N, ending this polling round.

In summary, MPLL reports all Twins pairs experiencing a state jumping in a short time interval with high probability.

Clustering the Twins pairs marked as ones having state jumping, we can sketch the region where the object stays. Minimizing the region helps to accurately localize the object. Ideally, the region that the object stays corresponds to a connected subgraph in G. In practice, it is possible that the moving target induces several unconnected subgraphs in G.

We tackle this problem by selecting the largest subgraph in G, termed as G_s. If there are more than two such subgraphs, we identify G_s as the minimum connected component in G containing those subgraphs, as illustrated in Fig. 4.1. After determining the possible region that the object stays, we use the centroid of the positions of all Twins in this region to approximate the object position.

4.2 Tracking

We use two algorithms to track the object, Kalman filter and particle filter.

4.2.1 Tracking the Object Based on Kalman Filter

Kalman filter algorithm was proposed by R.E.Kalman in 1960. It is an optimization recursive filter, also known as an autoregression filter. It can estimate the state of dynamic system via a series of measurements observed over time.

We treat the detection on moving objects as the series of measurements on the observation. Suppose the object is moving in a two-dimensional plane. Without loss of generality, the origin of coordinates is set as the entrance of the area under surveillance. We establish the rectangular coordinate system accordingly, and the variable of moving state for the object can be denoted as $X_t = \begin{bmatrix} x_t & y_t & v_{x_t} & v_{y_t} \end{bmatrix}^T$ ($t=0, 1, 2, \ldots k$), where the (x_t, y_t) is the coordinate of the object, v_{x_t} and v_{y_t} are the rectangular components of movement on the x and y axis, respectively. We assume that the sampling interval using the passive reader is one unit of time. The moving model of the target can be represented as

$$
\begin{cases}
x_k = x_{k-1} + v_{x_{k-1}} + a_{x_{k-1}} / 2 \\
y_k = y_{k-1} + v_{y_{k-1}} + a_{y_{k-1}} / 2 \\
v_{x_k} = v_{x_{k-1}} + a_{x_{k-1}} \\
v_{y_k} = v_{y_{k-1}} + a_{y_{k-1}}
\end{cases}
\tag{4.1}
$$

Where $a_{x_{k-1}}$ and $a_{y_{k-1}}$ are the acceleration components along the x and y axis following a $N(0, \sigma_a^2)$ distribution, respectively. The process noise w_k can be written as $\begin{bmatrix} 0.5a_{x_{k-1}} & 0.5a_{y_{k-1}} & a_{x_{k-1}} & a_{y_{k-1}} \end{bmatrix}^T$

Assume that the matrix for control model B_k is 0, the system state transition model is

$$
X_k = F_k X_{k-1} + w_k \quad w_k \sim N(0, \sigma_a^2 A) \tag{4.2}
$$

where

$$
F_k =
\begin{bmatrix}
1 & 0 & 1 & 0 \\
0 & 1 & 0 & 1 \\
0 & 0 & 1 & 0 \\
0 & 0 & 0 & 1
\end{bmatrix}
\tag{4.3}
$$

$$
A =
\begin{bmatrix}
0.25 & 0 & 0.5 & 0 \\
0 & 0.25 & 0 & 0.5 \\
0.5 & 0 & 1 & 0 \\
0 & 0.5 & 0 & 1
\end{bmatrix}
\tag{4.4}
$$

Utilizing the method as described in Sect. 4.1.2, we can obtain the observed value on the approximated position of the object y_k at a certain time point k. Then the observation matrix H_k is represented as:

$$
H_k =
\begin{bmatrix}
1 & 0 & 0 & 0 \\
0 & 1 & 0 & 0
\end{bmatrix}
\tag{4.5}
$$

Therefore, the procedure of Kalman filter based tracking algorithm is summarized as follows.

a. Initialization. Setting the initial state is $X_0 = [0, 0, 1, 1]^T$. Here the two former components, with the value 0, represent the entrance of the area. The two latter ones, with the value 1, represent the velocity components along the axis x and y of the moving object. The unit is m/s.

Fig. 4.2 Kalman filter
procedure

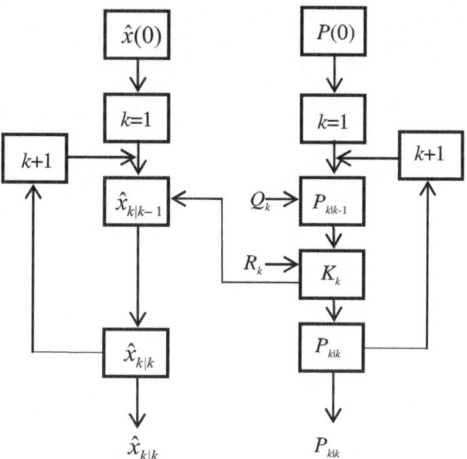

b. Prediction. Predicting the state of the object at time point k by using the one at the time point $k-1$, according to Eqn. 4.2.
c. Computing predicted estimate covariance matrix.
d. Computing the Kalman gain based on the observation at time point k. The Kalman gain is a function of the relative certainty of the measurements and current state estimate.
e. Updating state estimate using the results obtained from step (c) and (d).
f. Updating the (a posteriori) estimate covariance matrix at time point k.
g. Let $k=k+1$, repeating steps (b)—(g).

The entire procedure is illustrated in Fig. 4.2, where the $\hat{x}(0)$ and $P(0)$ represent the initial value of state estimate and filtered estimate covariance.

4.2.2   Tracking the Object Based on Particle Filter

The tracking procedure can also be completed via particle filter. With the approximated locations, we can use the particle filter method to identify the corresponding vectors in G to enable a timely tracking. We borrow the idea of particle filter to introduce a group of "particles" as random samples in the state space. Using those particles and observations at a specific time, we explore the distribution of a latent variable, i.e. the possible location of the moving object. Iteratively introducing particles and resampling is like filtering accurate estimations on the location and hence plot the trace for the object, as illustrated in Fig. 4.3.

When the object moves into the interrogation area, it will trigger some state jumping on some Twins pairs. Without loss of generality, we assume the start of the trace is at the main entrance of the area, where is also the origin of coordinates.

Suppose that $n_0 \sim n_8$ are the time stamps of state jumping at the corresponding Twins in Δt. The particle filter is performed via offline training and online detecting

Fig. 4.3 Particle filter based
tracking

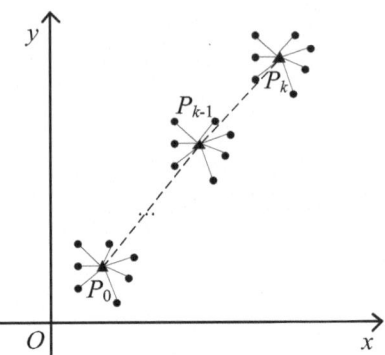

phases. In the former phase, we estimate the position of object l $_{P(n_1,n_2,\ldots,n_N|l)=\prod\limits_{i=1}^{N}P(n_i|l)}$.
The estimation discretizes the distribution P into histogram bins, and forms the
fingerprint of the position l.

In the second phase, the particle filter is conducted by performing the following
steps.

a. Initialization. Suppose the object moves from the origin of coordinates, with a
   speed as $v=<x_v, y_v>$, where x_v and y_v are the X-component and Y-component
   of speed v, respectively. We introduce N particles as samples near the origin of
   coordinates.
b. Prediction. The $p(X_k|Z_{1:k-1})$ is computed from the filtering distribution
   $p(X_{k-1}|Z_{1:k-1})$ at time $k-1$.
c. Weight computation. With the position of each particle. We calculate the weight
   based on two elements, the observation obtained by the measurement as well as
   the probability distribution of the time stamps of state jumping at each particle's
   position, resulting the weights as the probability of yielding the observation at
   each particle.
d. Re-sampling. We use the multinomial resampling [1] to remove the particles
   with low weights with high probability.
e. Approximation. With the position and weight of each particle, we can approximate the coordinates for the object at time k.
f. Let $k=k+1$, and iteratively perform the steps (b)—(f).

4.3   Experiment and Evaluation

We first investigate the key parameters for practical deployments. We then evaluate
the performance of Twins in real implementations.

4.3.1 Experiment Setup

We setup a Twins prototype system based on current commercial products. The system is fully compatible with existing standards, e.g. the EPCglobal UHF Class 1 Gen 2/ISO 18000-6C air protocol, and has no need to modify both the reader and tag. The system includes an Impinj Speedway R420 and 500 E41-b tags, which have been widely used in industry. The reader has a wide range of transmission power, from 10 to 32.5 dBm, and operates within the spectrum of 920~928 MHz.

We first calibrate the system for determining proper settings of Twins, as shown in Fig. 3.8. We check the performance of Twins in the localization and tracking. Core parameters include the distance between two tags in the Twins d, the angle θ between the antennas of Twins and the reader, the distance between the Twins and moving object D, and height from the Twins to the floor h.

We also implement the prototype in real inventory environments. We deploy our system on a number of aligned shelves. We set the distance between two shelves as 2 m and the distance of two adjacent Twin pairs as 0.6 m. We invite a volunteer to walk among the shelves. The performance in terms of detection rate is evaluated for probing the key settings for practical deployment.

4.3.2 Performance Evaluation

We vary the value of different settings to check change of successful detection rate r when localizing the moving object.

Normally the lobe width of the Impinj reader's antenna is 70°. Thus, we check the value of r by varying θ with 0°, 15° and 30°. For each θ, we change the distance D to 75, 105, 135, and 165 cm, respectively. With each distance D, the distance d is set as 6, 8, 10, 12, and 15 mm. Totally, we test 3*4*5=60 cases. We conduct 100 tests for each case. The calibration experiment is conducted as illustrated in Fig. 4.4.

The result summarized in Table 4.1 indicates that if $d=10$ mm the r is maximum in most cases. We set the default value of d as 10 mm in our implementation.

Fig. 4.4 Calibration for Twins deployment

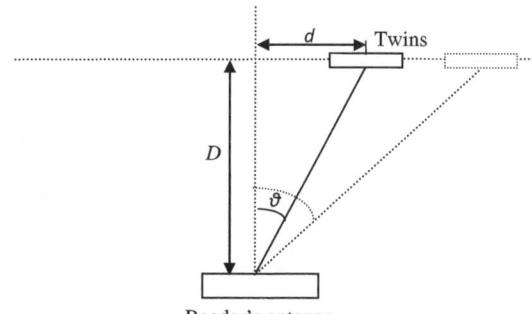

Reader's antenna

Table 4.1 Calibration for Twins deployment

	0°			15°			30°		
	min	max	ave	min	max	ave	min	max	ave
6 mm	93	100	98.25	73	93	83	98	100	99.33
8 mm	91	100	96	88	99	94	43	100	83.25
10 mm	97	100	98.5	96	98	97.25	89	98	95.2
12 mm	98	100	99.5	85	99	91.5	55	100	89.5
15 mm	96	100	98.5	81	99	93.25	70	100	92.6

We check the reliability of Twins when changing the distance between Twins and reader antenna D. We exam the transmission power P_{TX} for forcing the Twins into the critical state when changing D. The result shown in Fig. 4.5 indicates that a larger D requires a higher P_{TX}. In practice, the maximum deployment distance of Twins is highly related to P_{TX}. We find that the maximum distance is 5.8 m with maximum P_{TX} of 32.5 dbm when we use the Impinj R420 reader.

We also determine the proper height of Twins h in real deployment. We deploy Twins with different height settings and report the r in Fig. 4.6. The X axis in Fig. 18 represents the distance between the volunteer and Twins. Twins can achieve the highest detection rate, i.e. 95.17% in average, when h equals to 75 cm. This indicates that a Twins with its height $h=75$ cm can collect more RF signals reflected from the moving object. In this height, the multipath effect in indoor environments and the movement of arms and legs together help to the RF reflection. Hence, the Twins is easy to be triggered when the object moves. The r will be reduced when increasing or decreasing the height of Twins from the setting 75 cm, leading to a smaller impact on the Twins and a lower detection rate.

We investigate the accuracy of Twins-based tracking scheme and compare it with two well-known RFID based device-free approaches, LANDMARC [2]. LANDMARC is active tag based system. The tags are deployed in a tag array. The distance between the nearest neighbors in a row or column is 1 m for LANDMARC. During tracking the simulated intruder, we investigate the distance from the estimated position to the real position. We take those records as the error for tracking and plot the result in Fig. 4.7. The result shows that Twins has a better tracking accuracy than LANDMARC. The error rate of Twins is always below 0.85 m, and 0.75 in average.

Fig. 4.5 D v.s. P_{TX}

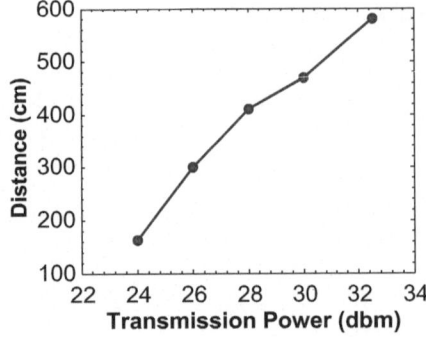

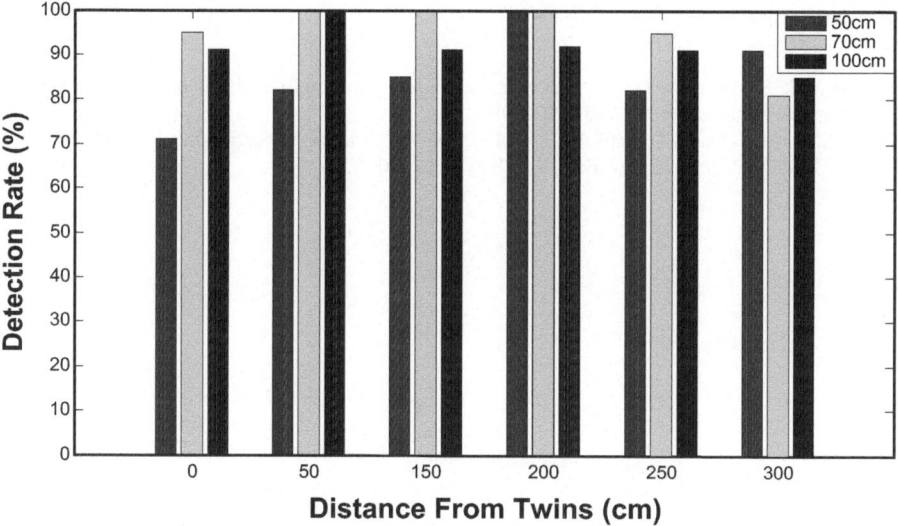

Fig. 4.6 The deployment height of Twins

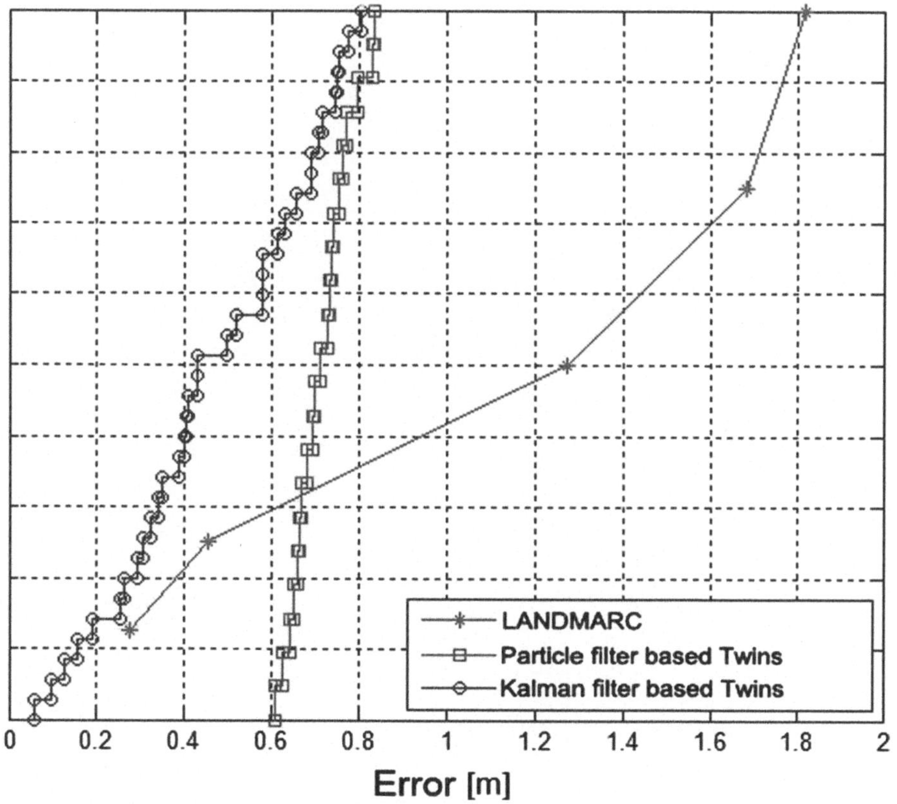

Fig. 4.7 The error during tracking the object

References

1. Pitt MK, Shephard N (1999) Filtering via Simulation: Auxiliary Particle Filters. American Statistical Association 94 (446):590–599
2. Ni LM, Liu Y, Lau YC, Patil A (2004) LANDMARC: Indoor Location Sensing Using Active RFID. ACM Wireless Networks, (WINET) 10 (6):701–710

Chapter 5
Sparse Deployment Based Tracking System Using Twins

The original Twins system has two limits for object tracking. First, Twins can only identify the location of an object but fail to provide the moving direction. In practice, the direction information is helpful to improve the localization and tracking accuracy. Second, the grid deployment pattern of Twins systems should fully cover the area of interest. Achieving fast and accurate tracking may require a dense deployment of Twins, resulting in relatively high cost.

We then design a Twins based Sparse Deployment (TSD) for tracking objects. TSD enables an accurate and efficient tracking while significantly reducing the number of Twins. More important, TSD is able to recognize both the direction and location of the object using sparsely deployed Twins system. TSD leverages the intersection of inventory or warehouse areas. In each intersection, TSD deploys three Twins pairs at the corner, as shown in Fig. 5.1.

In such a layout, we deploy a few Twins pairs at the critical points, including the main entrance, exit, and cross points of aisles, realizing a sparse deployment of Twins. Meanwhile, at each critical point, the Twins pairs can report the moving direction of the target object through the signal analysis. In this way, we only use a few passive tags to track the object in the device-free pattern.

5.1 Fundamentals of Direction Indication

The main challenge (and contribution) of our passive RFID based system is to accurately predict the direction of a moving object in real-time. Addressing this issue significantly reduces the deployment cost and increases the detection efficiency.

The fundamental of the proposed direction indication algorithm is based upon both the observations from experiments and theoretical modelling of backscatter communication. As aforementioned, we define the *critical power* (CP) of a tag as the power of the transmission (reader) with which the tag can be turned into its critical state. We find that for per-location of an intruder, monotonic relationship between the CP of deployed tags and the position of the object does not hold. However, for

© The Author(s) 2014
J. Han et al., *Device-Free Object Tracking Using Passive Tags,* SpringerBriefs
in Electrical and Computer Engineering, DOI 10.1007/978-3-319-12646-3_5

Fig. 5.1 Deployment
scenario

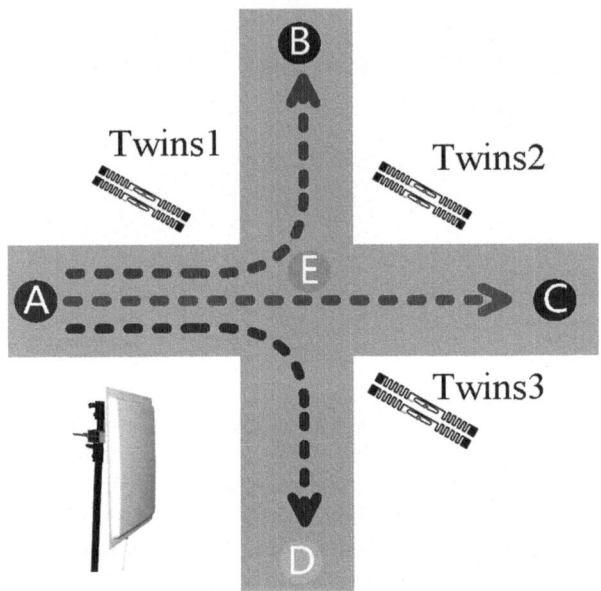

whole-trajectory in each monitoring intersection, a unique correspondence between
the features of CP sequence and trajectory holds well, i.e., the eigenvalues of the CP
matrix can be used as an indicator for plotting an intruder's trajectory.

In the rest of this subsection, we detail the theoretical analysis supporting the
above claims, and present the verification result of indoor experiments.

5.1.1   Impact of Object Location to Received Power

We establish the functional relationship between the average power density and
the object's location. According to the Poynting theorem [1], the Poynting vector
S, which is interpreted as an instantaneous power density, can be calculated as the
cross product of electric intensity **E** and magnetic intensity **H**, i.e.,

$$\mathbf{S} = \mathbf{E} \times \mathbf{H} \tag{5.1}$$

on the other hand, according to the Maxwell's equations [1], we have $\mathbf{H} = \mathbf{a_s} \times \dfrac{\mathbf{E}}{\eta}$ if
the wave is Uniformed Plain Wave, where $\eta \approx 120 \ \pi\,\Omega$ is the wave impedance. $\mathbf{a_s}$ is
a unit vector of the Poynting vector **S**.

Theorem 1 If the environment variables are constant and the wave satisfies the Uniformed Plain Wave (UPW) [2] condition, the tag's receiving power is determined by
the moving object's location.

Proof The wireless channel between the transmitter antenna and receiver antenna can be modeled as a linear time-varying (LTV) system [3]. Therefore, the electric intensity can be separated into two parts: the stable environmental effect and the moving object's effect.

Environmental Effect Assuming the reader's transmit intensity is $\mathbf{E_R}$, the received intensity of the tag in stable scenario $\mathbf{E_T^C}$ can be denoted as [4].

$$\mathbf{E_T^C} = \mathbf{E_R} \sum_{i=0}^{m} L_{RT_i} e^{j(\alpha - \varphi_i)} \tag{5.2}$$

where m, L_{RT_i}, φ_i and j are the number of distinguishable paths, the path loss of the ith path from reader to the tag (including the surface reflection), the corresponding time delay and the imaginary unit, respectively.

According to the phasor formula [5], we have

$$\mathbf{E_T^C} = \mathbf{E_R} \mathfrak{L}_{RT} e^{j(\alpha - \varphi_{RT})} \tag{5.3}$$

and

$$\begin{cases} \mathfrak{L}_{RT} = \sum_{i=0}^{m} L_{RTi}^2 + 2 \prod_{0 \le i < j \le m} L_{RT_i} L_{RT_j} \cos\left(\varphi_i - \varphi_j\right) \\ \\ \varphi_{RT} = \arctan \dfrac{\displaystyle\sum_{i=0}^{m} L_{RT_i} \sin \varphi_i}{\displaystyle\sum_{i=0}^{m} L_{RT_i} \cos \varphi_i} \end{cases} \tag{5.4}$$

where $\mathfrak{L}_{RT}$ and φ_{RT} denote the path loss and phase delay contributed by all multipath components respectively.

Moving Object Effect When an object P enters into this area, it changes the electromagnetic field distribution. The intensity contributed by the moving object [3] is

$$\mathbf{E_T^P} = \mathbf{E_R} \mathfrak{L}_p e^{j(\alpha - \varphi_P)} \tag{5.5}$$

where $\mathfrak{L}_P$ and φ_P denote the path loss and phase delay contributed by the intrusive object.

Due to the linearity of LTV, the joint intensity $\mathbf{E_T}$ can be calculated as

$$\begin{aligned} \mathbf{E_T} &= \mathbf{E_T^C} + \mathbf{E_T^P} \\ &= \mathbf{E_R} \left(\mathfrak{L}_{RT} e^{j(\alpha - \varphi_{RT})} + \mathfrak{L}_p e^{j \cos(\alpha - \varphi_P)} \right) \end{aligned} \tag{5.6}$$

Substituting Eq. 5.6 into Eq. 5.1, we have

$$\begin{aligned}
\mathbf{S} &= \mathbf{E} \times \mathbf{H} \\
&= \mathbf{E_R} \times \mathbf{a}_\eta \times \mathbf{E_R} \frac{1}{\eta} \left(\mathfrak{L}_{RT} e^{j(\omega t - \varphi_{RT})} + \mathfrak{L}_p e^{j\cos(\omega t - \varphi_P)} \right)^2 \\
&= \frac{E_R^2}{\eta} \left(\mathfrak{L}_{RT} e^{j(\omega t - \varphi_{RT})} + \mathfrak{L}_p e^{j\cos(\omega t - \varphi_P)} \right)^2 \mathbf{a}_S
\end{aligned}$$

Therefore, the time-average power density is

$$\begin{aligned}
\langle \mathbf{S} \rangle &= \mathbf{a}_S \frac{E_R^2}{\eta T} \int_0^T \left(\mathfrak{L}_{RT} e^{j(\omega t - \varphi_{RT})} + \mathfrak{L}_p e^{j\cos(\omega t - \varphi_P)} \right)^2 dt \\
&= \mathbf{a}_S \frac{E_R^2}{2\eta} \left(\mathfrak{L}_{RT}^2 + 2\mathfrak{L}_{RT}\mathfrak{L}_p \cdot \cos\left(\varphi_P - \varphi_{RT}\right) + \mathfrak{L}_P^2 \right)
\end{aligned}$$

Assuming the coordinates of the reader, tag and object are $R = (X_R, Y_R)$, $T = (X_T, Y_T)$ and $P = (X_P, Y_P)$, respectively. The length of trajectory $\overline{RPT}$ from the reader to tag via the object is $d_{RPT} = \sqrt{(X_P - X_R)^2 + (Y_P - Y_R)^2} + \sqrt{(X_T - X_P)^2 + (Y_T - Y_P)^2}$.

The path loss on $\overline{PRT}$ can be modeled as

$$\mathfrak{L}_P = \frac{\alpha}{d_{RPT}} \tag{5.7}$$

where α is a proportionality constant [2]. Due to the half-wave loss phenomenon [1], the equivalent phase delay is

$$\varphi_P = 2\pi \frac{d_{RPT}}{\lambda} - \pi \tag{5.8}$$

Therefore, we can see that the norm of average power density $\langle \mathbf{S} \rangle$, denoted as $\langle S \rangle$, is determined by the transmitting power density $\frac{E_R^2}{2\eta}$, the environmental parameters $\mathfrak{L}_{RT}$ and φ_{RT}, the parameters $\mathfrak{L}_P$ and φ_P. $\mathfrak{L}_P$ and φ_P are correlated with the moving object's position, as shown in Eqs. 5.7 and 5.8. In other words, if the environment remains unchanged, the received power of the tag is only determined by the object's location, i.e.

$$\langle S \rangle = f(\mathfrak{L}_P, \varphi_P) = f(P) \tag{5.9}$$

It is worth noting that a power density $\langle S \rangle$ is not one-to-one correlated to a unique location P, due to the non-monotonic relationship. Thus, we are aiming to extract a power sequence uniquely mapping to a trajectory of the moving target.

5.1.2 Weak Correlation Between Trajectory and Power Sequence

Denote a path sequence $Path_1 = \langle p_1, p_2, \cdots, p_m \rangle$, and the corresponding power sequence as

$$S_1 = \langle s_1, s_2, \cdots, s_m \rangle = \langle f(p_1), f(p_2), \cdots, f(p_m) \rangle$$

where f is defined by Eq. 5.9. We repeat n samples, because the volunteers cannot step each position so exactly. Suppose the offset of ith sample of position p_j is $\Delta p_j^i = p_j^i - p_j$. Therefore, for the ith sample, the path sequence is $Path_1^i = \langle p_1^i, p_2^i, \cdots, p_m^i \rangle$, and the corresponding power sequence is $S_1^i = \langle f(p_1^i), f(p_2^i), \cdots, f(p_m^i) \rangle$. Because f is derivable, using the Taylor formula, we have

$$f(p_j^i) = f(p_j + \Delta p_j^i) \approx f(p_j) + \Delta p_j^i \frac{df}{dp} \tag{5.10}$$

For ease of expression, we use p express the position instead of the coordinates (x, y). Based on the 1st order differential invariant, if the coordinates of p is (x, y), then

$$\Delta p_j^i \frac{df}{dp} = \Delta x_j^i \frac{\partial f}{\partial x} + \Delta y_j^i \frac{\partial f}{\partial y},$$

Where Δx_j^i and Δy_j^i are the offsets from the coordinates x_j^i and y_j^i, respectively. Therefore, the observed power matrix can be

$$S^o = S^a + W$$

$$= \begin{bmatrix} f(p_1^1) & f(p_2^1) & \cdots & f(p_m^1) \\ f(p_1^2) & f(p_2^2) & \cdots & f(p_m^2) \\ \vdots & \vdots & \ddots & \vdots \\ f(p_1^n) & f(p_2^n) & \cdots & f(p_m^n) \end{bmatrix} + W$$

$$\overset{(5-10)}{\approx} \begin{bmatrix} f(p_1) & f(p_2) & \cdots & f(p_m) \\ f(p_1) & f(p_2) & \cdots & f(p_m) \\ \vdots & \vdots & \ddots & \vdots \\ f(p_1) & f(p_2) & \cdots & f(p_m) \end{bmatrix} + \frac{df}{dp} \begin{bmatrix} \Delta p_1^1 & \Delta p_2^1 & \cdots & \Delta p_m^1 \\ \Delta p_1^2 & \Delta p_2^2 & \cdots & \Delta p_m^2 \\ \vdots & \vdots & \ddots & \vdots \\ \Delta p_1^n & \Delta p_2^n & \cdots & \Delta p_m^n \end{bmatrix} + W$$

$$= F + \frac{df}{dp} \Delta P + W$$

Because W is independent to the power and the offset Δp is independent to the position p, the covariance matrix of S^o is

$$\mathbf{cov}(S^o) = E\left((S^o)^H S^o\right)$$

$$= E(F^H F) + \frac{df}{dp} I \sigma_{\Delta P} + I \sigma_W$$

$$= U\left(\Sigma_F + \frac{df}{dp} I \sigma_{\Delta P} + I \sigma_W\right) U^H \qquad (5.11)$$

Now, we use another power sequence S_{new} (the corresponding path sequence is $Path_{new} = \langle p_{new1}, p_{nwe2}, \cdots, p_{newm} \rangle$) instead of the last row of S^o. If the $Path_{new}$ is quite different from $Path_1$, the 1st order Taylor formula of $Path_1$ cannot be used to approximate $Path_{new}$.

Therefore, the S^o_{new} is

$$S^o_{new} = S^a_{new} + W$$

$$= \begin{bmatrix} f(p_1^1) & f(p_2^1) & \cdots & f(p_m^1) \\ f(p_1^2) & f(p_2^2) & \cdots & f(p_m^2) \\ \vdots & \vdots & \ddots & \vdots \\ f(p_{new1}^n) & f(p_{new2}^n) & \cdots & f(p_{newm}^n) \end{bmatrix} + W$$

$$\overset{(5-10)}{\approx} \begin{bmatrix} f(p_1) & f(p_2) & \cdots & f(p_m) \\ f(p_1) & f(p_2) & \cdots & f(p_m) \\ \vdots & \vdots & \ddots & \vdots \\ f(p_{new1}^n) & f(p_{new2}^n) & \cdots & f(p_{newm}^n) \end{bmatrix}$$

$$+ \frac{df}{dp} \begin{bmatrix} \Delta p_1^1 & \Delta p_2^1 & \cdots & \Delta p_m^1 \\ \Delta p_1^2 & \Delta p_2^2 & \cdots & \Delta p_m^2 \\ \vdots & \vdots & \ddots & \vdots \\ 0 & 0 & \cdots & 0 \end{bmatrix} + W$$

$$= F_{new} + \frac{df}{dp} \Delta P' + W$$

The covariance matrix of S^o_{new} is

$$\mathbf{cov}(S_{new}^{o}) = E\left(\left(S_{new}^{o} \right)^{H} S_{new}^{o} \right)$$

$$= E\left(F_{new}^{H} F_{new} \right) + I\,\sigma_{\Delta P'} + I\,\sigma_{W}$$

$$= U_{new}\left(\Sigma_{F_{new}} + I\,\sigma_{\Delta P'} + I\,\sigma_{W} \right) U_{new}^{H} \tag{5.12}$$

5.1.3 Experimental Validation for Theoretic Analysis

We conduct indoor experiments to verify the properties from theoretical analysis.

Figure 5.2 shows the CP sequences with different trajectories. There are three tags deployed in an intersection. The different colors denote different values of CPs in the samples. The former two sub-figures show the CP variations with the same trajectory in two runs, while the latter two plot the results with the other trajectory. It shows that using the absolute values of their CPs for device free position estimation is not reasonable since identical CP values may correspond to different positions. However, for each individual trace, CP values have a relatively stable changing mode. In addition to the one-CP multiple-position issue, there are many other factors affecting CP based sensing. In particular, unpredictable environmental factors are very common. For example, the radio signal path loss is unknown and costly to profile in most cases since it is temporally dynamic and spatially unevenly distributed.

As confirmed by our empirical data, the relationship between a trace of the object and the CP sequence reported by tags is stable. Hence, collecting such sequences and extracting their features is possible to identify the motion direction of objects.

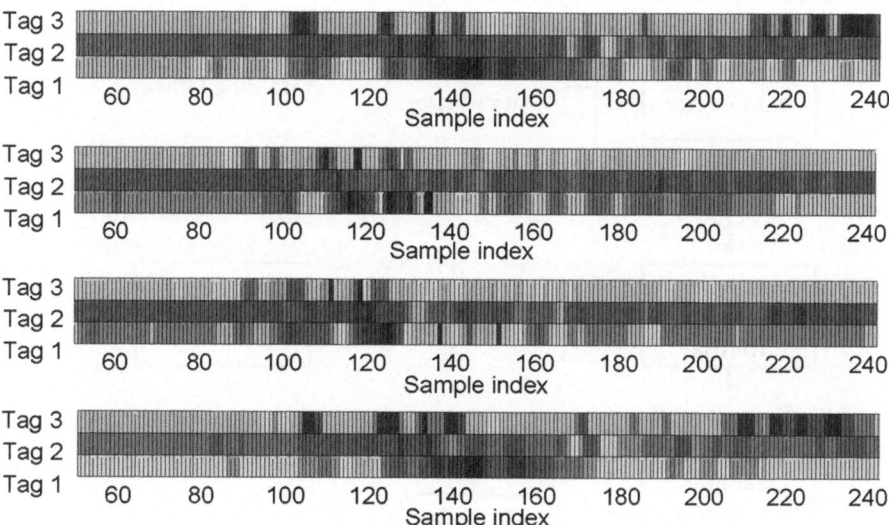

Fig. 5.2 CP variance with different moving trajectories

This heuristic correlation has not been disclosed by previous localization and motion detection approaches using passive tags. They merely provide a "0/1" judgment indicating whether the object is detected or not, resulting in a degradation in both the accuracy and time-efficiency.

5.2   System Design

Based on the theoretical analysis in Sect. 5.1.2, we propose a device-free method to recognize and predict an intruder's movement direction. As shown in Fig. 5.3, our algorithm consists of two phases, data preprocessing and trajectory recognition. The algorithm 5.1 generates a ternary sequence with a certain length from the raw data. The algorithm 5.2 identifies the moving direction through the extracted motion profiling.

5.2.1   Data Preprocessing

In the data preprocessing step, there are three stages. Here we still use the critical power. The critical power (CP) of a tag is transmission power of reader which just

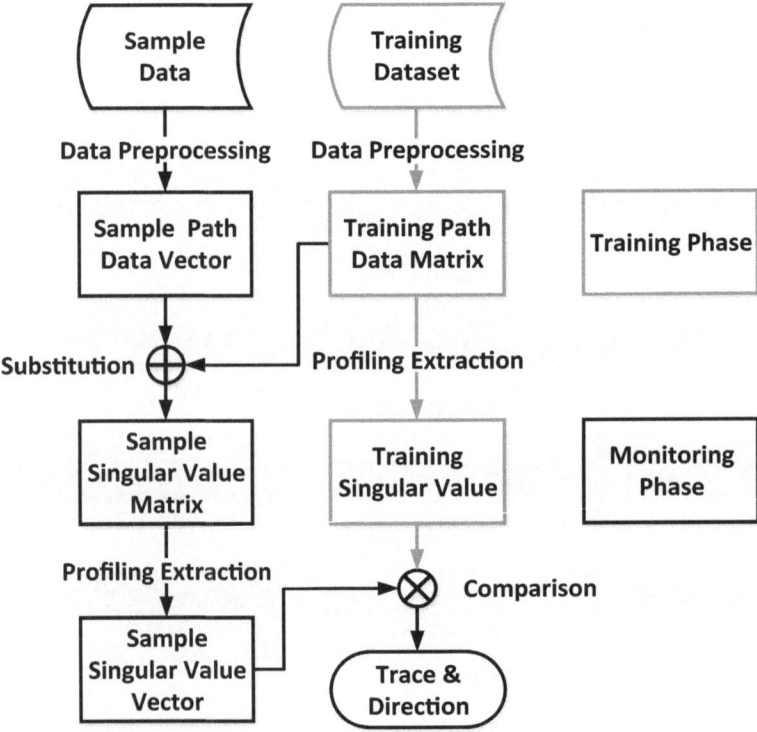

Fig. 5.3 Flow chat of TSD

forces the tag into its critical state. In the first stage, the change of a tag's CP is a trinary number c (i.e. -1, 0, 1), which means a tag's CP increases, keeps unchanged, and decreases, respectively. As motioned in Sect. 5.1.2, objects at different locations will have different effects on a tag's CP. In some locations, the object reflects RF signals, which will inject more RF waves to the tag such that it can be read with a smaller CP, where $c=1$. In some locations, the object blocks the transmission from the reader to tag, weakening the power received by the tag such that it should be read with a larger CP, where $c=-1$. In some locations, the object has no impact on the tag's CP, where $c=0$.

The reader continuously scans the CP of each tag and obtains a sequence of its trinary c. In order to react to an intruder efficiently, the reader should filter abundant runs of 0s out, because there is not an intrusion in those runs. The details of truncation method will be discussed in Sect. 5.3.

Additionally, the length of trinary number sequence varies when the object moves at different speeds. In order to obtain the same length of sequence N in the training step and monitoring step, we normalize the sequence using the linear mapping.

Algorithm 5-1: Data preprocessing

Input: $rawData; \mathcal{L}; \mathcal{U}; \mathcal{N}$
Output: $pathData$

1 $L = length(rawData);$
2 **for** $i = 1 : L$ **do**
3   **for** $tagId = 1 : 3$ **do**
4   **if** $rawData(i, tagId).cp < \mathcal{L}$ **then**
5              $temp(i, tagId) = 1;$
6   **end**
7   **else**
8   **if** $rawData(i, tagId).cp < \mathcal{U}$ **then**
9                  $temp(i, tagId) = 0;$
10   **end**
11   **end**
12   **else**
13             $temp(i, tagId) = -1;$
14   **end**
15   **end**
16   **if** $temp(i, :) = 0$ **then**
17          $temp(i, :) = [];$
18   **end**
19 **end**
20 $truncate(temp);$
21 $K = length(temp);$
22 **for** $i = 1 : K$ **do**
23      $pathData\left(\left[\frac{N-1}{K-1}i + \frac{K-N}{K-1}\right]\right) = tempData(i);$
24 **end**

5.2.2   *Motion Profile Extraction and Tendency Recognition*

As discussed in Sect. 5.2.1, we know that introducing a different path will change the eigenvalue and the eigenvector of the covariance matrix. We use U as the motion profiling. For a matrix M, because the eigenvectors matrix of its covariance matrix $cov(M) = E(M^H M)$ is equal to the singular vector matrix of M [6], we utilize the Singular Value Decomposition (SVD) instead.

Finally, the movement trace and direction are recognized by comparing the motion profiling in monitoring step with that in the training step.

We divide the moving trace into 12 kinds for each intersection as shown in Fig. 5.1.

In the training step, we repeat T times measurement of trinary sequence S_{ij}^o for each kind of the movement trace P_j, and then extract the motion profiling U_j for trace P_j by Algorithm 5.2.

Algorithm 5-2: Path identification

    Input: $S_1^o, S_2^o, \cdots, S_T^o; S'^o; stepFlag$
    Output: p
1 **if** $stepFlag == ``Training"$ **then**
2   **for** $j = 1 : pathId$ **do**
3        $U(j) = \mathrm{SVD}(S_1^o, S_2^o, \cdots, S_T^o);$
4   **end**
5 **end**
6 **else**
7   **for** $j = 1 : pathId$ **do**
8        $U'(j) = \mathrm{SVD}(S'^o, S_2^o, \cdots, S_T^o);$
9        $D(j) = \|U'(j) - U(j)\|;$
10   **if** $D(j) < minDis$ **then**
11         $minDis = D(j);$
12         $p = j;$
13   **end**
14   **end**
15 **end**

In the monitoring step, we can obtain a trinary sequence S'^o. In order to identify the actual movement trace, we replace S_{1j}^o with S'^o to obtain the motion profiling $U'j$, and then calculate the distance D_j between U_j and $U'j$. For each kind of trace P_j, we can get a motion profiling distance D_j. The trace P_j that corresponds to the shortest distance D_j is the actual movement trace.

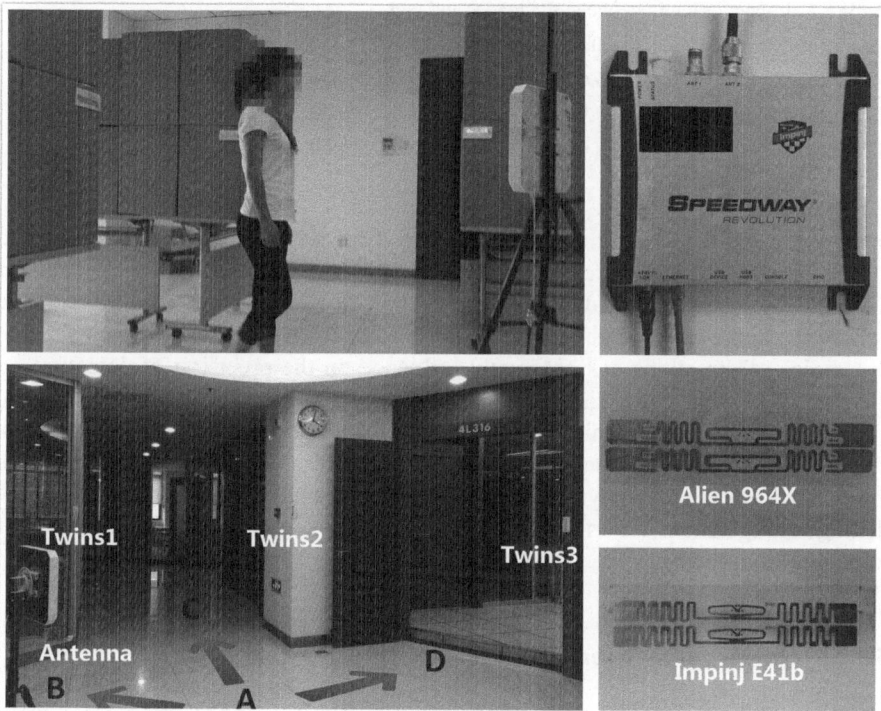

Fig. 5.4 Experimental setup

5.3   Experiments and Evaluations

5.3.1   Experiment Setup

We conduct experiments in a warehouse area to evaluate the tracking performance of TSD. As shown in Fig. 5.4, We use several Impinj SPEEDWAY 220 readers. The antenna we used is Laird A9028R30NF with an 8 dbi gain, 10–32.5 dBm transmission power, and 920~928 MHz working spectrum. We select three commodity passive tags, i.e. Impinj E41b, E41-c, and Alien 964x, in our experiments, which have been widely used in logistics and inventory systems.

5.3.2   Performance Evaluation

5.3.2.1   Impact of Different Angles

As shown in Table 5.1, we use the connecting line between the central antenna of reader and Twins2 as the benchmarks, with an interval in every 5°. To investigate

Table 5.1 Impact of angle between the reader and tags

Angle	−15° (%)	−10° (%)	−5° (%)	0° (%)	5° (%)	10° (%)
Average	83.33	96.67	90.00	90.00	83.33	83.33
Upbound	90.00	100.00	100.00	100.00	100.00	90.00
Lowbound	70.00	90.00	80.00	80.00	70.00	70.00

the accuracy of detection, the antenna is turned from 15° anti-clockwise (−15°) to 10° clockwise (10°). The reason why we drop other directions is that Twins1 or Twins3 may become unreadable while the angle between Twins2 and the antenna is over the angle referred above due to the directions of antennas. We can see that the choice of directions has obviously influence on the precision of detection. In the case of 10° anti-clockwise (−10°), the rates of detection accuracy are 96.67, 90 and 100 %, respectively, which is much higher than other comparative experiments. The symmetry may take the responsibility for the result of 0° because at the moment, Line AB is almost symmetric to Line AD, leading to certain impact on the distinguishing ability. The effect made by varying the angle on the accuracy appears to be non-asymmetry.

5.3.2.2   Impact of Different Heights

We fix the altitude of the reader antenna by 1 m over the ground. The center of antenna is parallel to the planes of twins tag. The twins are tied to the shells with different height for the use of testing the accuracy under variant directions. We can see that when the tag is at the same altitude of the antenna, we get the best performance. As shown in Table 5.2, in the 6 groups of experiments, the accuracy of 4 groups has reached 100 % and only one group is no less than 80 %, which is much higher than the accuracy at other altitude. The main reasons to obtain such results are that the effective area of the tags becomes largest and the SNR reaches the highest level. Note that when the tags are approaching to the ground, there are more outliers in the experimental data. This is incurred by the serious noisy wave and the low quality of single due to the reflections from the ground. While the tags are close to the celling, the power of reflected signals is relatively lower because of a longer distance between the tags and ceiling.

Table 5.2 Impact of tag height

Height	0.6 m (%)	0.8 m (%)	1.0 m (%)	1.2 m (%)	1.4 m (%)	1.6 m (%)
Average	95	92	98	90	92	90
Upbound	100	100	100	100	100	100
Lowbound	80	80	80	60	70	70

Fig. 5.5 Impact of intersection width

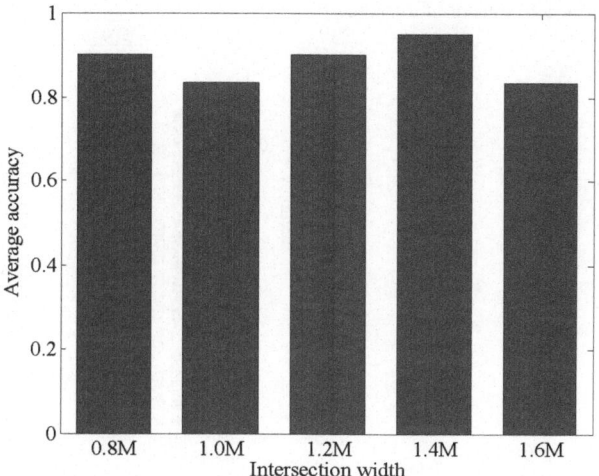

5.3.2.3   Impact of Intersection Widths

In Fig. 5.5, we investigate the accuracy of the direction recognition in a variety of crossing width. We used paper boxes to simulate the crossing aisles with different interaction width, ranging from 0.8 to 1.6 m. Little difference in performance appeared in the case of 0.8 m, 1.2 m and 1.4 m and the average accuracy reached over 90 %. The performance of 1 m case is relatively little worse for the reason of fading when the signals are transmitted to the reader via different paths in the multipath environment. In the case of 1.6 m, the accuracy declined obviously. Because besides the effect of multipath, when the crossing aisles is very wide, even if the movements are in the same direction, there may be great difference in critical power, resulting in an accuracy decline.

5.3.2.4   Impact of Different Kinds of Tags

In Fig. 5.6, we plot the performance of three kinds of tags in TSD. As we can see, Impinj E41c produces the best performance. The accuracy rates in three directions are classified with 90, 100 and 90 % respectively. The performance of Alien 964x and E41b is relatively lower, but the worst one is still over 80 % and the average is 90 % which may be related to the quality of tags. The tags of Alien 964 series are more suitable for consumer electronic devices mobile asset tracking. However, it is not wise to apply tags to the surface of wood, plastic, paper and glass. E41b is designed to provide a long read range for low-dielectric-constant materials. E41c is more suitable for high-dielectric-constant surfaces. Typical applications include tagging of carton and plastic boxes, apparel, books and file folders.

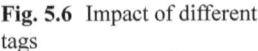

Fig. 5.6 Impact of different tags

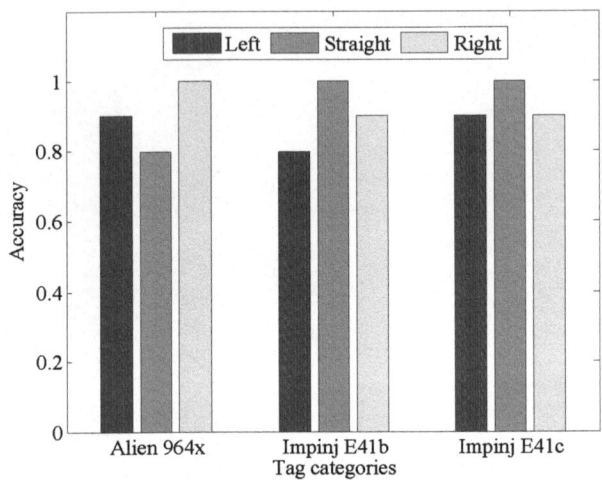

5.3.2.5   Impact of Moving Velocities

In the same scenario, we evaluate the impact of moving speed on the tracking accuracy. Volunteers walk through the intersection many times at different speeds, e.g. the slow (1.5 m/s), middle (2 m/s), and fast (2.5 m/s) speed. Figure 5.7 depicts the accuracy of the direction recognition at different velocity of movements. TSD achieves the best accuracy, i.e. 96.7 % in average, when the target moves at the middle speed. And the accuracy of detecting low-speed objects is about 93.3 % in average, while for the high speed, the accuracy is about 83 % in average.

For an object passing the intersection with different speeds, system will obtain different sequence lengths of CP. For easy figures, Algorithm 5.1 normalizes different number of sampling points into the same number of samplings. Since the medium speed is used in the training phase, the similar speed has best accuracy in

Fig. 5.7 Impact of different velocities

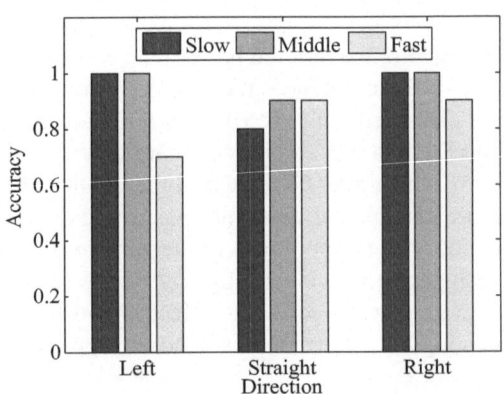

Fig. 5.8 Impact of different people

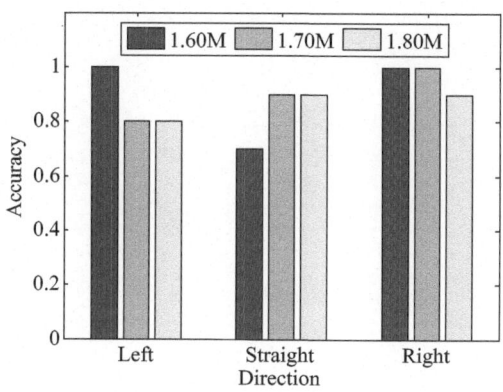

monitoring phase. In low speed experiments, it can be regarded as oversampling, which also has a good performance. In contrast, the high speed experiments will have rarer samplings, which has a relative lower accuracy than medium speed and low speed.

5.3.2.6 Impact of Different People

People in different height have a certain impact on the accuracy of direction recognition, as shown in Fig. 5.8. We find that the accuracy variance of taller persons in three directions is relatively lower. A taller person has a lager radio receiving and reflection area. Thus, the characteristics of his movement in three directions are more likely to significantly facilitate the accurate direction recognition.

References

1. Hayt WH, Buck JA (2001) Engineering Electromagnetics, McGraw-Hill.
2. Cheng DK (1989) Field and wave electromagnetics, vol 2. Addison-Wesley.
3. Tse D, Viswanath P (2005) Fundamentals of Wireless Communication. Cambridge university Press.
4. Franceschetti G, Stornelli S (2006) Wireless Networks: From the Physical Layer to Communication, Computing, Sensing and Control. Academic Press.
5. Serway R, Jewett J (2013) Physics for Scientists and Engineers. Cengage Learning.
6. Leon SJ (1980) Linear Algebra with Applications. Macmillan Press.

Afterword

This book has focused almost exclusively on using passive RFID tags and readers for device free tracking.

We have only touched on the variety of passive RFID tags based device-free user tracking and complete protocols that are suitable for densely or sparsely deployed environments. We have still left for others some important issues, such as balancing the tradeoff between deployment density and tracking accuracy, optimally scaling the system to support multi-users tracking, and integrating the proposed techniques into enterprise services. We have also left the important security issues unexamined, such as the system availability analysis, protocol design verification, attacking models, and etc.

I hope the reader found this book informative, and perhaps on occasion useful for both the research and development purposes. I'd like to offer thanks again to the many people who helped, and accept responsibility for the inevitable errors of commission and omission. Comments, corrections, and criticism may be addressed to us at {hanjinsong, weixi.cs, and zhaokun2012} @mail.xjtu.edu.cn.

September 14, 2014

Jinsong Han
Wei Xi
Zhiping Jiang
Kun Zhao

© The Author(s) 2014
J. Han et al., *Device-Free Object Tracking Using Passive Tags,* SpringerBriefs
in Electrical and Computer Engineering, DOI 10.1007/978-3-319-12646-3

Index